EXPLOR

THE WORLD OF ASTRONOMY

FROM THE CENTER OF THE SUN TO THE EDGE OF THE UNIVERSE

JOHN HUDSON TINER

First printing: August 2013
Sixth printing: April 2022

Master Books® is a division of the New Leaf Publishing Group, Inc.

ISBN: 978-0-89051-787-1
ISBN: 978-1-61458-354-7 (digital)
Library of Congress Number: 2013943397

Cover by Diana Bogardus

Unless otherwise noted, Scripture quotations are from the New Intervational Version of the Bible.

Please consider requesting that a copy of this volume be purchased by your local library system.

Printed in the United States of America

Please visit our website for other great titles:
www.masterbooks.com

For information regarding author interviews, please contact the publicity department at (870) 438-5288.

This book is dedicated to Ivan Wyatt Tiner

Photo Credits: T=Top, B=Bottom, L=Left, R=Right
All images from shutterstock.com unless stated.
California Institute of Technology, Courtesy of the Archives: 103T
ESO: 110, 111, 114, 139, 143, 150
Library of Congress: 68, 135,
Luc Viatour: 7, 12
NASA: 4, 9, 14, 19T, 46, 55, 71, 75B, 79, 82B, 83, 98, 100L, 102, 104, 112, 119, 121, 122, 128R, 140, 147, 149, 171, 172
NASA ESA: 3, 127BR, 134
NASA/JPL-Caltech: 20, 25, 26BR, 27, 28, 31TR, 32, 36, 38, 39, 40, 44, 45, 51, 52, 57, 58, 59, 60, 62, 63, 64, 69, 70, 72, 73, 74, 75T, 81, 82T, 84, 85, 113, 125, 128L, 146
NOAA: 48
Public Domain: 8, 23, 43, 49, 55, 56, 80, 87R, 93, 94, 95, 96, 97B, 99T, 101, 107, 127TL, 127BL
Royal Astronomical Society / Science Source: 100R
Science Source: 86L
United States Naval Observatory: 24, 37BL
Wikipedia: 10, 11, 13, 19B, 23, 26TL, 26BL, 31TL, 31B, 33, 37TL, 37R, 48, 50, 67, 86, 88, 89, 90, 94, 97T, 99B, 110, 115, 116, 123, 124, 127TR, 133, 144

Master Books®
A Division of New Leaf Publishing Group
www.masterbooks.com

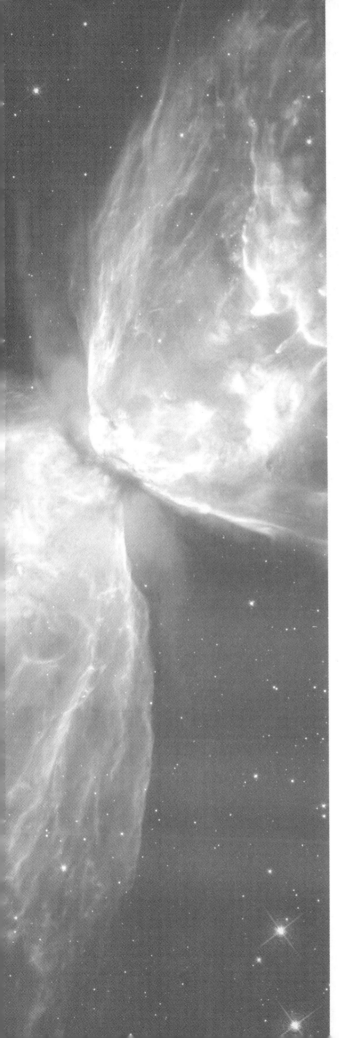

The Butterfly Nebula

Table of Contents

Note to Parents and Teachers:
How to use *Exploring the World of Astronomy*

Students of several different ages and skill
levels can use *Exploring the World of Astronomy*.
Children in elementary grades can enjoy many
of the concepts, especially if given parental
help. Middle school students can read the book
independently and quickly test their understanding
and comprehension by the challenge of answering
questions at the end of each chapter. Junior high and
high school students can revisit the book as a refresher
course. They will gain greater insight by using the
section below the questions marked "Explore More."
The suggestions in "Explore More" offers questions,
discussion ideas, and research for students to develop a
greater understanding of astronomy. Additional "Explore
More" opportunities start on page 162 and a whole book
review can be found on page 165.

A Hubble Telescope image of the barred
spiral galaxy NGC1300, located almost
69 million light years from Earth

Exploring the Moon

Astronomy is basically the study of the night sky — the moon, planets, stars, and deep space objects. Whether pursued as a profession or a hobby, astronomy is a fascinating subject filled with unexpected discoveries and delightful sights. This book will look at all aspects of observational astronomy and will offer examples of how personally to learn more about the night sky with nothing more than the unaided eyes, binoculars, and a small telescope.

As a form of recreation, astronomy can be enjoyed with simple equipment. The unaided eyes are capable of wide-angle views and of rapid movement that cannot be matched by any other optical instruments. The eyes can follow the swift passage of shooting stars. The eyes can scan the Milky Way and search out star patterns. Only with the eyes is it possible to see constellations in their entirety — the regal head of Leo the Lion and his hunched hindquarters, the symmetry of the twins of Gemini, the quick curl

Explore

1. What is the moon's terminator?

2. What was the Great Moon Hoax?

3. What is a blue moon?

Progress

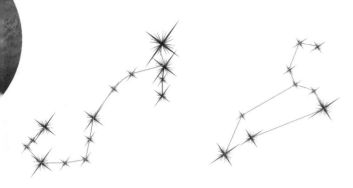

The constellations of the Scorpion, Leo, and Gemini

of the Scorpion's stinging tail to his red heart. The Milky Way, star groupings, meteorites, comets, and other subjects can be seen and explored by merely going out on a dark night and looking to the heavens.

Even the moon is a satisfying subject to first explore with the eyes alone.

The moon in its cradle of stars is the favorite object of beginning astronomers. Even experienced professionals cannot resist viewing the moon time and again. Galileo said, "It is the most beautiful and delightful sight." The moon is a rugged and grand place. It has sharp peaks, dust-covered plains, and ring-shaped craters. Mountains reflect light like crinkled aluminum foil. Vast dark plains, called *maria* (singular *mare*), cover part of its surface.

The moon has many interesting features visible to the eyes alone. At first, the moon's surface may seem a baffling confusion of craters, seas, and mountains. But once an outstanding feature is found, it can serve as a guide to other sights. Such a guidepost is Mare Crisium — the Sea of Crises.

Here's the way to find Mare Crisium. Look at the moon when it rises in the east. Imagine the moon as a clock face with twelve o'clock at the top. Let your eye travel along the curve of the disk. At about two o'clock you will find a small, dark oval.

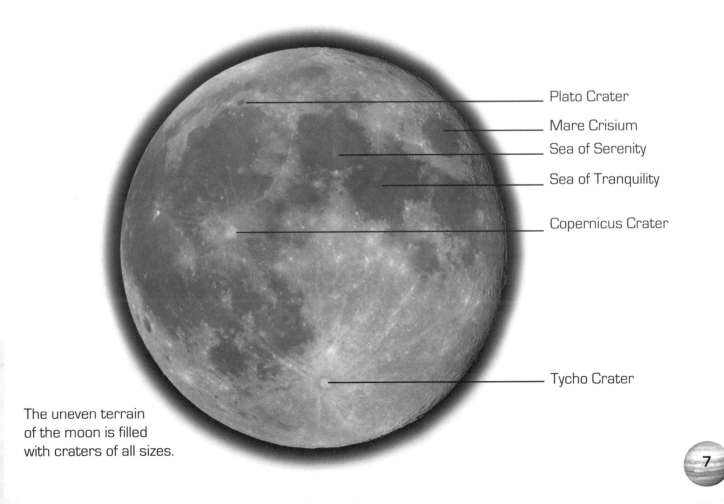

Plato Crater

Mare Crisium

Sea of Serenity

Sea of Tranquility

Copernicus Crater

Tycho Crater

The uneven terrain of the moon is filled with craters of all sizes.

Anyone with normal eyesight can find the spot quite easily. The dark oval is Mare Crisium, one of the many seas of the moon. It is a great plain, three hundred miles in diameter, surrounded on all sides by mountains.

Mare is Latin for sea, but in the case of the moon, the maria are bleak plains of lava. The moon is a harsh body. The temperature soars to more than 250°F during the day, then plunges to a couple hundred degrees below zero at night. In addition, the moon has no atmosphere. Astronauts who walk upon its surface wear space suits both as protection from heat and cold and to also provide oxygen to breathe. The first people on the moon were Neil Armstrong and Buzz Aldrin, United States astronauts. On July 20, 1969, they landed on the Sea of Tranquility.

You can find the Sea of Tranquility very easily with the unaided eyes. Use Mare Crisium as a guide. The Sea of Tranquility is the large, dark plain located halfway along a line connecting Mare Crisium to the center of the moon. The Sea of Tranquility is larger than Mare Crisium, but irregular in shape.

Not only is the Sea of Tranquility the first place astronauts landed, but also it is the site of the first church service on the moon. Because they landed on Sunday, Astronaut Aldrin thought it appropriate to hold a worship service to God. In the one-sixth gravity he read a verse from the Bible, said a short prayer, and took communion.

Above the Sea of Tranquility is the Sea of Serenity. It can be seen with the unaided eyes, too. When the line between light and dark cuts across

Galileo's telescopes

the sea of Serenity it looks like a giant crater. High mountains rim it all around. The range of mountains circling the Sea of Serenity can just be seen with the unaided eyes under the right conditions. Look when the moon is half full (first or third quarter moon). The mountains curve like a silver sword cutting into the dark side of the moon.

Shadows are as important as light to reveal small features on the moon. Shadows make details more easily seen. Watch for smaller features when they fall along the terminator — the line dividing light from dark. Near the terminator, shadows give a contrast that makes details more easily seen.

When Galileo turned his first telescope to the moon, he was astonished at the number of craters. They crowded over each other, spilled across each other, and even fell inside one another. Craters of all sizes are sprinkled everywhere across the moon, even in the maria.

Ordinary binoculars are as powerful as Galileo's first telescope. They are powerful enough to reveal craters, including three that are easy to find. Tycho

Galileo's drawing of the moon

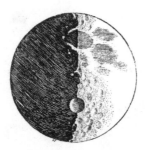

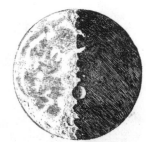

Crater near the lower edge of the moon is named after a famous astronomer, Tycho Brahe. The crater has long white rays spreading out from it. The white streamers can be seen running halfway across the moon.

Astronomers believe the white rays are material thrown out as the craters formed due to an impact by a large meteor or comet. Over time, small meteors bombard the rays, and the moon's intense temperature changed their color. The rays faded away. If this idea is true, then Tycho is a relatively young crater.

Copernicus Crater, named after the Polish astronomer, is found near the center of the moon. It is a blaze of white. Compare it with the inky black of the third well-known crater, Plato, which is at the top of the moon. Plato is named after one of the great ancient Greek thinkers.

Once you find these craters with binoculars, you will be able to glimpse them with the unaided eye. They are best seen when the line dividing light from dark passes through them. The longer shadows make the crater walls stand out better. One rim is thrust into bright sunlight, while the other rim is still inky black in darkness.

Watch from night to night as the moon changes phases and the line dividing light from dark marches across the face of the moon. The inky blackness of new moon changes to the bleached raw whites of full moon. Mountain peaks catch the light first. With binoculars, you can spot the jagged peaks of the mountains. In a telescope, the peaks seem to float like pyramids of flame because of the bright sunlight. See if you can spot the speckle of a peak poking out of darkness into light.

Despite its smaller size, the moon's mountains are as tall as those on earth. Galileo measured one of the mountains on the moon to be four miles high. When the light is right, a high mountain can have shadows one hundred miles long. It seems strange that we can look up and see the mountains of another world hanging over our heads.

In the 1960s, the United States prepared for astronauts to land on the moon. Before the moon shot, the United States sent unmanned vehicles to test the moon's surface. Cameras would transmit photos, and instruments would test the mineral content of its surface. The first unmanned probe sent to land on the moon was *Surveyor 1*.

Some scientists expressed concern that the moon's surface would not support a vehicle. Without an atmosphere, meteorites that impacted its surface

The uneven terrain of the moon is filled with craters of all sizes.

The Great Moon Hoax

One of the most outlandish and famous hoaxes of all time involved the moon. The word hoax means to trick people into believing as true something that is false. The moon has no seas of water, although early astronomers did not know this. The fact that the features were named *maria*, meaning seas, caused some people to think there might be life on the moon. In 1835, a New York newspaper took advantage of this interest in the moon to boost sales of the paper.

The *New York Sun* newspaper printed a series of reports about life on the moon. According to the *Sun*, the famous British astronomer Sir John Herschel had invented a new type of telescope. John Herschel was the son of William Herschel who discovered the planet Uranus. Like his father, he was an accomplished and well-respected astronomer. The newspaper claimed that John Herschel turned his powerful telescope to the moon. He saw beaches, trees, animals and birds. Finally, he saw bat-like moon people.

John Herschel could not be questioned. He was on an expedition to South Africa. The *Sun* newspaper claimed to base its stories upon articles that appeared in the *Edinburgh Journal of Science*, a Scottish scientific magazine.

Great Moon Hoax lithograph of "ruby amphitheater" for *New York Sun*, August 28, 1835

Other newspapers could not find the Scottish journal. Rather than questioning the astonishing story, they pretended to have access to the original articles in the journal. They simply rewrote the *Sun* articles and passed them off as their own as if they had independently verified the story.

Finally, after many months, John Herschel was reached in South Africa. He had no super telescope, nor had he written about the moon. The Scottish journal did not exist. Instead, a reporter for the *Sun* had written the stories to boost newspaper circulation. The *Sun* enjoyed the largest circulation of any newspaper in the world at that time.

would have turned bedrock into powder. The extremes of heat and cold would have done the same thing. The best, smoothest landing sites would be the ones most likely covered by powder. Some scientists feared *Surveyor 1* would land in a sea of soft powder. It would be swallowed up as if by quicksand when it touched down.

On June 2, 1966, *Surveyor 1* touched down on the moon. The landing pads sank only a couple of inches into the loose surface, hardly more than it would have on earth. The probe's camera sent back 11,150 photographs before it shut down for the long night on the moon.

The flat areas of the moon are covered with a layer of fine particles. It is as much as 5 to 20 feet deep in some places. However, the particles stick together. They make a firm surface.

The moon is earth's only natural satellite. The moon is unusual because of its large size. Its diameter, 2,160 miles across, is a little more than one-fourth the diameter of the earth. Our moon is unusually large compared to the planet it revolves around. When seen from another planet such as Mars, it and the earth would appear as a double planet, unlike anything else in the solar system.

Mercury has no moon and neither does Venus. Mars has two satellites, but they are exceptionally small, being about 5 and 10 miles in diameter. They would easily fit in some of the smaller craters on the moon. The four gas giant planets — Jupiter, Saturn, Uranus, and Neptune — have many satellites. Three of the satellites of Jupiter, and one of the satellites of Saturn are larger than the moon. But Jupiter and Saturn are big planets and about 20 times larger than their satellites. Overall, the moon is the fifth largest satellite in the solar system.

The sun is 400 times the diameter of the moon, but it is also 400 times farther away. For that reason, the moon appears to be about the same size as the sun when viewed from the earth. The moon can exactly blot out the light of the sun. No other moon can do that. The moons of Mars are too small and only take a small bite out of the sun. The moons of Jupiter and the other more distant planets are too far away from the sun to cover it despite their larger size.

An eclipse of the sun can only take place during new moon, when the moon is between the earth and the sun. The moon blocks part or all of the sun's light. An eclipse does not take place every new moon. The tilt of the moon's orbit explains why it does not do so. It passes above or below the sun.

The earth's orbit takes it around the sun, and the surface containing the orbit is known as the ecliptic plane. If the moon's orbit

were also in the ecliptic plane, then there would be a solar eclipse every month. However, the moon's orbit is slanted by about five degrees. The tilt sends it above the ecliptic plane for part of its orbit and below the ecliptic plane for the rest of the orbit. Eclipses of the sun do occur fairly often somewhere on earth, but not every month.

A solar eclipse occurs when the shadow of the moon is cast on the earth. People inside the shadow have the light of the sun blocked. The shadow often misses blocking the sun entirely because its orbit may be slightly above or below the ecliptic. The moon blocks part of the sun. It looks like the moon took a bite out of the sun. The result is a partial solar eclipse.

The two other types of eclipses are total and annular. Neither the earth nor the moon's orbits are perfectly circular. They are elliptical. The path along an elliptical orbit changes how far the earth is from the sun, and how far the moon is from the earth. On occasion, the earth and the moon are farther from the sun, and the moon is not quite large enough to completely cover the sun as seen from

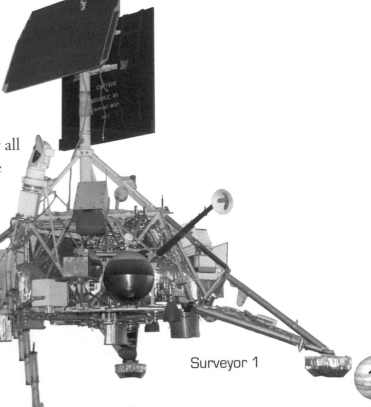

Surveyor 1

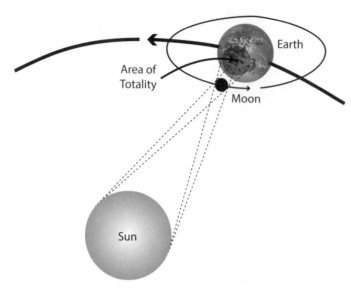

earth. The result is a ring of sunlight around the moon. The eclipse is an annular eclipse. *Annular* means ring.

When conditions are right, however, the moon's apparent size is large enough to completely blot out the sun — a total solar eclipse. A total solar eclipse is a dramatic event. Daylight slowly fades, the land becomes dusty looking, wind blows uncertainly, and the temperature drops. Birds become confused and fly away to find a place to roost for the night. Suddenly all sunlight fades. Brighter stars shine in a purple sky.

The earth's rotation limits how long the eclipse lasts. The moon's shadow sweeps across the rotating earth. In any one location, a total eclipse lasts only about five minutes. The longest time possible is seven minutes, and no eclipses of that length will occur during all of the years from 2000 to 2100.

Often, a solar eclipse is only visible in out of the way places, or over the ocean. In any one location, it

may be 400 years between a total eclipse. Amateur astronomers, and professionals, too, fly halfway around the globe to be in the right place to see a total eclipse.

In ancient times, when people did not travel, an eclipse was a frightful event. To avoid a panic, kings hired skilled astronomers to calculate the times of eclipses to forewarn the people.

The earth's moon is large enough and close enough to affect the earth itself. It can produce tides. The satellites around other planets are too small, or the planet is too large for the satellite to affect their surface. Not the earth. Our moon can produce tides. (The sun has a role, too.)

Anybody who has been to the ocean has marveled at an incoming tide. They listen to the rhythm of the pounding surf, feel the cool shower of spray, and step back as waves crash against rocks.

A total solar eclipse in progress

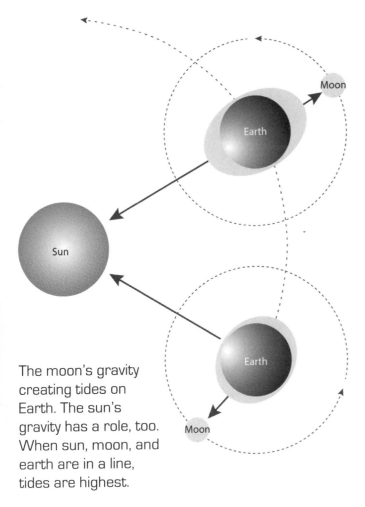

The moon's gravity creating tides on Earth. The sun's gravity has a role, too. When sun, moon, and earth are in a line, tides are highest.

Canada's Bay of Fundy when the tide is high (above) and when the tide is low (below)

Isaac Newton discovered the cause of tides. The moon's gravity attracts two heaps of water that seem to move as the earth rotates. Twice each day the water rises for six hours and then ebbs for the next six hours. The sun enters the picture, too. Highest high tide and lowest low tide occur one after another when sun, moon, and earth are in a straight line.

Along the United States coastline tides average about four feet. The land it strikes controls the size of a tide. Some bays are long and sloping with funnel-shaped shores. The shores tighten together and squeeze water into smaller bays. Tides in the Bay of Fundy on the northeast coast of Canada can reach 50 feet or more. In the open ocean where there is no land to cause the water to pile up, tides are about two feet. Lake Michigan has tides of two inches.

If you watch the moon from one night to the next night, you will discover two facts. First, it rises later each night. Second, the moon's appearance changes. The moon goes through phases that depend upon the position of the sun and moon.

You will also notice that you always see the same side of the moon. It rotates on its axis and revolves around the earth in such a way to always keep the same side facing the earth. The earth's gravity has locked onto the moon and prevents it from looking away.

You may have heard the expression "the dark side of the moon." In this case, the word "dark" means hidden from view and unknown. Before space probes orbited the moon, astronomers did not know what the back side of the moon looked like. The features of the far side of the moon were a mystery. In 1959, the Soviet spacecraft *Luna-3* returned the first images of the hidden side

The back, or "dark," side of the moon does receive sunlight, but its features were unknown until spacecraft orbited the moon.

has set. The illuminated portion that we can see from earth grows each night. The moon goes from new to waxing crescent (*waxing* means to increase slowly in size), then first quarter (half full), waxing gibbous, full moon, waning gibbous (*waning* means to decrease slowly in size), last quarter, waning crescent, then new moon again.

The size of a crescent moon is measured by how many hours have passed since new moon. A one-day-old crescent moon is one day past new. It becomes visible low in the west just as the sun sets. The crescent is extremely thin. Sometimes the rest of the unlighted portion is faintly visible. It is illuminated from sunlight that strikes the earth, reflects out into space, shines on the moon, and reflects back to earth. The faint light is called earthglow.

The moon goes around the earth in a little over 27 days, 7 hours. However, during that time, the earth continues ahead in its travels around the sun, carrying the moon with it. Because sunlight causes the phases of the moon, the moon has to travel a

of the moon. It proved to have craters, mountains, and lava plains similar to the visible side, although not as rough overall.

The "dark" side of the moon may be hidden from view, but it does receive sunlight. Half of the moon is always in sunlight, just as half of the earth has day while the other half has night. Only during a full moon, when the side facing earth is fully lighted, is the back side fully plunged in darkness. The rest of the time at least some of the back side of the moon has its surface in sunlight.

The phases of the moon depend on how much of its lighted side is visible from earth. When the moon is between the sun and the earth, the lighted side faces away from earth, and the unlighted side faces us. It rises with the sun and sets with the sun. This phase is the new moon. The word "new" means a fresh start.

About two days after new moon, a thin crescent moon appears shortly after sunset in the west near where the sun

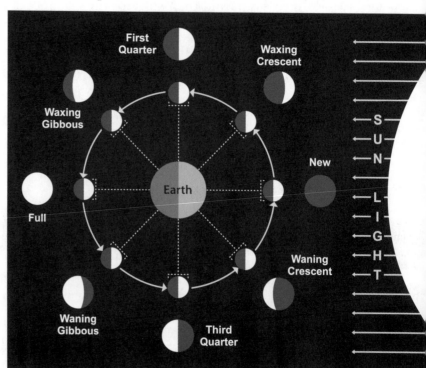

Different phases of the moon

little farther to get in the right position to start the next series of phases. The moon needs about 29 days, 12 hours to go from new moon to new moon again.

The description of the phases of the moon can be misleading. The first quarter moon is not one-fourth full. Instead, it is half full. "Quarter" refers to the fact that the moon is one-quarter of its way through its complete cycle of phases.

On average, the moon rises 50 minutes later each night. However, the delay in rising changes throughout the year. In the fall, the earth's daylight hours and nighttime hours are equal, but the moon's orbit is tilted. It is higher in the northern hemisphere at the start of fall, which makes the delay in its rise from one day to the next to be only 30 minutes.

The full moon nearest to the first day of autumn (about September 23) rises in the east just as the sun sets in the west. The next night it rises only 30 minutes later and continues to do so for several nights. Before twilight ends, the moon provides its glow. Farmers welcomed the extra illumination of the full moon because they can continue to bring in their crops after the sun has set. They called it the harvest moon.

The next full moon is similar to a harvest moon. It is known as the hunter's moon.

Usually, only one full moon occurs in each month. But with a full moon occurring every 29.5 days, it goes through 12 complete cycles in 364 days — but the year has 365 days (and leap year has 366 days). Eventually, a year gains an extra full moon, and two full moons occur in the same month. A second full moon in the same month is called a blue moon. A blue moon is, of course, the same color as any other full moon. The expression "once in a blue moon" means that something is rare. Smoke from a forest fire can cause the moon to have a dusty, blue color. That seldom happens — it is a rare occurrence — and such a rare blue moon due to smoke gives its name to the extra moon in a month.

The table "Moon Facts" provides a summary of the data about the moon. The average distance of the moon from the earth is about 240,000 miles, about 10 times around the earth at the equator. This is a small distance as astronomical distances go, and even as ordinary distances. For instance, a big rig trucker will drive his 18-wheeler about half that distance in a year. Some cars such as taxis will put more than 240,000 miles on them before they are worn out. The perigee distance is the nearest the moon comes to the earth, and the apogee is the farthest.

Orbital period, about 27.25 earth days, is the time for the moon to go around the earth once. A day on the moon, about 29.5 earth days, is the time from one sunrise on the moon to the next sunrise on the moon. A day on the moon is also the time from new moon to new moon.

The moon is about 0.27 the diameter of earth, or slightly more than one-fourth. Diameter can be misleading because surface area increases by the square of the diameter and volume by the cube of the diameter. The earth has a surface area 13 times greater than the moon's surface area. The continent of Asia has about the same surface area as the

Table of Moon Facts	
Average distance from earth	238,854 miles
Perigee	225,291 miles
Apogee	251,910 miles
Orbital period	27 days, 7 hours, 43 minutes
Day	29 days, 12 hours, 44 minutes
Diameter	2,160 miles
Mass	1/81 the earth's mass
Gravity	0.17 earth normal (1/6th)
Density	3.3 g/cm³
Albedo	0.14
Day temperature	261° F
Night temperature	-279° F
Magnetic field	none

moon. The earth has a volume 50 times the volume of the moon.

Mass is a measure of the amount of matter in an object. The earth is 81 times more massive than the moon. If balance scales could be built to weigh them, it would take 81 moons on one side to balance the weight of the earth on the other side.

Gravity is a measure of the force of attraction. If we set the force of gravity on the surface of the earth equal to 1.0, then the force of gravity on the moon is about 0.17, or 1/6 the earth's gravity. A person who can lift 100 pounds on earth would be able to lift 600 pounds on the moon. A person could hit an object farther, too. Astronaut Alan Shepard of the Apollo 14 moon mission attached a golf club head to the handle of a sampling arm and hit two golf balls. If he could hit them at the same speed as on earth, they would go about a mile, maybe farther without air resistance. But Shepard was hampered by thick gloves and an awkward spacesuit. He had to use one arm to swing at the ball. No one knows how far the ball went. But most people agree he holds the record for the longest golf shot.

Density is a measure of mass compared to volume. Water has a density of 1.0 grams per cubic centimeter. The earth has a density 5.5 times that of water, and the moon has a density of 3.3 times that of water. A representative sample of moon material would weigh less than a similarly sized sample of earth material.

Albedo is how well the surface reflects light. The word is Latin for "whiteness." A perfectly reflective surface would have an albedo of 100 percent (1.00). A surface that absorbs all the light that hits it would have an albedo of 0 percent (0.00). The earth has an albedo of about 30 percent (0.30), although that can change depending upon cloud cover, snow, and time of year. The moon's albedo is 14 percent (0.14). It is a dark object and absorbs 86 percent of the light that falls on it.

The moon lacks both an atmosphere and water to moderate its temperature. The moon's surface temperature swings from scorching hot in the day to bitterly cold at night.

The moon also has no magnetic field. A compass could not be used to give direction on the moon.

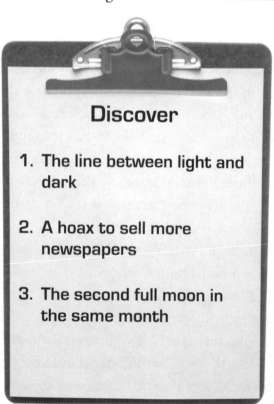

Discover

1. The line between light and dark

2. A hoax to sell more newspapers

3. The second full moon in the same month

T F   1. The only features on the moon visible with the eyes alone are the maria.

A B C D   2. The seas of the moon are made of _____.
              (A. dried up lake beds B. lava C. soot-covered ice D. vast deserts)

T F   3. The first moon landing by astronauts was in the Sea of Tranquility.

A B   4. The United States spacecraft *Surveyor 1* (A. sank out of sight in a deep dust layer on the moon B. successfully landed and took several thousand photographs).

A B C D   5. The diameter of the moon is a little more than _____ the diameter of the earth.
              (A. one-tenth B. one-fourth C. one-half D. twice)

T F   6. The moon's tallest mountains are about one-fourth the size of those on earth.

T F   7. An eclipse of the sun can only take place when the moon is between the earth and the sun.

_____   8. The three types of solar eclipses are partial, annular, and _____.

A B   9. The earth's moon is unusually _____ compared to the planet it revolves around.
              (A. large B. tiny)

A B C D   10. The time from new moon to new moon is _____.
              (A. 24 hours B. seven days C. 29.5 days D. 365.24 days)

_____   11. The full moon nearest to the first day of autumn is called the _____ moon.

T F   12. The average distance of the moon from the earth is about 240,000 miles.

EXPLORE MORE

From a map of the moon, select a feature named after a person, research his or her life, and write a two-page report that summarizes the individual's achievements.

What is the difference between a solar eclipse and a lunar eclipse?

Sketch the features that you can see when the moon is half full and again when it is full. What type of features did you see better under each phase?

Observe the full moon rising and later in the night. Some people say the moon appears larger when on the horizon than when overhead. What causes this optical illusion?

For the Math Whiz: Calculate the speed of the moon in its orbit around earth. Speed = distance/time. Distance is the circumference of the orbit given by $C = 2\pi r$, with r the moon's distance from earth. Time is the number of hours in a complete orbit. Refer to the table of moon facts for the needed information.

Mars

Ancient stargazers did not have to contend with streetlights or the glow of large cities. When the sun set and twilight faded, the sky became a black expanse filled with a glittering spray of stars. Those who kept lonely vigils at night — shepherds, hunters, and sailors — became well acquainted with the night sky. The star patterns became easy to remember because they did not change. As the stars marched across the sky because of the earth's rotation, they remained locked in position with one another.

However, five objects moved against the background of fixed stars. Ancient Greeks called the objects *aster planetes,* meaning "wandering stars." The word *pla-nets* means "wandering," and the word *aster* means "star." It is also found in asteroid (star-like) and asterisk (the * symbol found on a keyboard). The five planets did move. They followed the same path across the sky as the moon and tracked through the same constellations.

At first, planets received names linked to their color. Venus shown with a

Explore

1. Why did Christiaan Huygens use an "aerial" telescope?

2. Why did so many astronomers turn telescopes to Mars in 1877?

3. Why were the moons of Mars so difficult to discover?

An image of Mars from the Hubble Space Telescope

The Bible relates that during the Apostle Paul's second missionary journey, he traveled to Athens, the chief city of Greece. He spoke to the learned men of Athens from the Areopagus, a word in Greek that means "Aries' Rock." Because Mars is the modern name for Aries, Aeropagus is sometimes translated as Mars' Hill as it is in the King James Version of the Bible. "When Paul stood in the midst of Mars' hill . . ." (Acts 17:22; KJV).

The first five planets can easily be seen with the unaided eyes, although not all are visible at the same time. Their orbits sometimes take them into the daytime sky and close to the sun. When in the night sky, they are easy to detect. All five are brighter than Sirius, the brightest star.

In addition to being so bright, planets move slightly against the background of stars from one night to the next. They shine with a steady light and seldom twinkle as stars do.

Why do stars twinkle but not planets? Stars are so distant they appear as points of light in the night sky. Rising heat currents cause turbulence in the air. Starlight shining through the unsteady air flickers. Even on nights with the air unusually clear and steady, the stars shine with a pinpoint brilliance

brilliant white; Mars a vivid red. About 700 B.C. the Greeks gave Venus the name Phosphorus — the light-bearing one. Mars earned the name Pyroesis — the fiery one. Later, Venus became Aphrodite, the Greek goddess of love. Mars became Ares, the Greek god of war. Its red color matched the rusty red of iron swords stained with blood.

The Romans renamed the planets after their mythological beings. Venus is the Roman name for the goddess of love. Ares became Mars, the Roman god of war. By about 150 B.C., the planets had received the names by which they are still known today: Mercury, Venus, Mars, Jupiter, and Saturn. The remaining two planets, Uranus and Neptune, had not yet been discovered.

Mars Hill, overlooking Athens, where the Apostle Paul preached

that is quite unlike the planets. Planets are far closer to the earth than the stars. They have a disk, not quite visible to the unaided eyes, but large enough to shine with a more steady light.

Mars is the fourth planet from the sun. It lies outside the orbit of earth. When it is closest to earth, the sun illuminates its full face, and it shines brighter than any star. Because of its distinctive color, Mars is often called the Red Planet. Its only rivals in color are Aldebaran, a red star in the constellation of Taurus the Bull and Antares in Scorpus the Scorpion. The name *Antares* in Greek means "anti-Ares," or "Rival of Mars."

In some ways, Mars is similar to earth. Mars spins on its axis once in 24 hours, 37 minutes, giving it a day only 37 minutes longer than an earth day. Its axis tilts 25 degrees, while earth tilts 23 degrees. The tilt gives Mars four seasons like the earth. But Mars is 1.5 times as far from the sun as the earth, so it has a colder climate. Mars' year — the time it takes to go around the sun — is 687 days. That is about twice as long as earth's year.

Mars is only 4,220 miles in diameter. It is the second-smallest planet — Mercury is smaller. (So is Pluto, but Pluto is no longer considered a major planet.) Mars has a surface area about equal to the dry land on earth.

Because of its small size, no additional detail can be seen in binoculars than can be seen with the eyes alone. Binoculars do enhance its color. Instead of uniform red, Mars takes on a deep reddish-yellow color. Mars often passes near other sky objects such as the moon, one of the other planets, or a bright star. Binoculars then make a nice sight with Mars and the moon or Mars and another planet in the same field of view.

In a telescope, the color becomes less red and more like butterscotch. Seeing detail on the Martian surface is a challenge for earth-based telescopes. The largest ones reveal features only about 200 miles across. It is an even more difficult subject for small telescopes.

The best time to examine Mars in any telescope is when it is opposite the earth from the sun. This close approach is known as opposition. When in opposition, Mars presents a fully lighted face to earth. Even during the most favorable condition, Mars is 70 times smaller than the full moon. Amateur astronomers with ordinary-size telescopes report seeing the white caps at the north and south poles of Mars. Sometimes they get faint glimpses of surface features.

Ancient astronomers observed two puzzling aspects of Mars. During its travels across the sky,

Mars would change in brightness. When well placed high in the sky at midnight, it would be far brighter than Sirius. Later, as it moved closer to the sun, it would grow dimmer than the top 25 brightest stars. Why did it change in brightness?

Its motion puzzled them, too. Normally, it traveled in a counterclockwise direction against the background of stars. But occasionally, just before it became its brightest, it would slow down, stop, and go in a backward direction, and then move forward again. Mars would travel in reverse for about 72 days before going forward again. What could explain the backward or retrograde motion?

Until the 1500s, most people, and astronomers, too, believed the sun, moon, and planets traveled around the earth. They put the earth at the center of the planetary system, and had everything else going around the earth in circular orbits. Ptolemy, a Greek scientist, lived about A.D. 100. He took the planetary observations of Hipparchus, who lived about 100 years earlier, and worked out the earth-centered system in mathematical detail.

In the earth-centered system, Mars would stay the same distance from the earth and would be equally bright during all its travels. Yet, during a period of two years Mars faded from a bright red object to a much dimmer one. Mars changed in brightness by a factor of about 75 — a big difference. The earth-centered planetary system offered no explanation.

But Ptolemy could predict with reasonably good success eclipses, alignments of the planets, and other celestial events. For the next 1,400 years, astronomers stayed with his system. However, successful results took long and tedious mathematical calculations.

Nicolas Copernicus looked for a better way. Copernicus was a Polish astronomer who lived in the 1500s. He found the math was easier if he switched the location of earth and sun. He put the sun at the center of the planetary system. The earth and planets traveled around the sun in circular orbits. His system made the calculations easier, but did not give any better results when predicting events. Both Ptolemy and Copernicus based their systems on the same old star charts of Hipparchus. The old charts gave inaccurate results by either method.

However, the sun at the center of the planetary system explained why Mars changed in brightness. Sometimes the orbits of earth and Mars put them near one another and Mars looked brighter. At

The surface of Mars is covered in many places with boulders and wind blown sand. Twin Peaks known as hummocks (small knolls) are shown in the background of an image by the Mars Pathfinder.

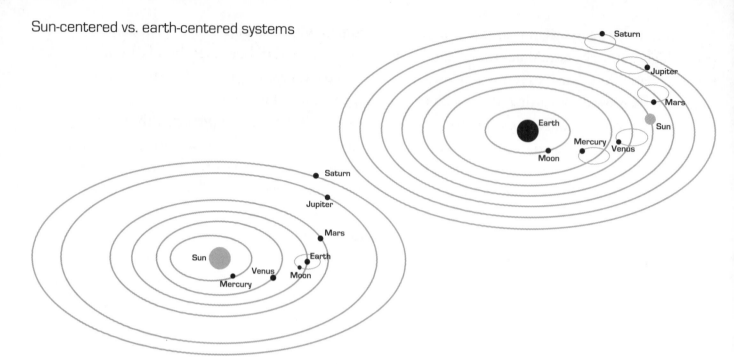

other times, Mars was on the other side of the sun from earth, and it became dimmer because of the greater distance.

Retrograde motion could be easily explained, too. Earth followed an orbit nearer the sun than Mars and traveled around the sun more quickly. Because of its speedier course, earth overtook Mars. For an observer on earth, Mars fell behind and appeared to travel in reverse.

Astronomers did not instantly accept the sun-centered planetary system. They based one objection on parallax. Parallax is the apparent movement of an object caused by the actual movement of the observer. You can demonstrate parallax by making a thumbs-up fist in front of your eyes at arm's length. If you close one eye, you will see the thumb against some object in the background. If you hold your thumb steady and close the other eye instead, you will see a change in your thumb's apparent position against the background. This change in position is parallax.

In the sun-centered planetary system, earth traveled around the sun. Because of this motion, astronomers should see the stars shift in position. That did not happen. Copernicus could only

suggest that the stars were so far away that parallax became so slight as to be unnoticed.

One hundred years passed before the idea that the earth traveled around the sun became more widely accepted. In 1609, astronomers tried to detect parallax with the newly invented telescope but once again failed. Galileo's telescope did not have the power to show parallax. Astronomers could only conclude, as Copernicus had, that the stars occupied a vast distance from the earth. Eventually, they would succeed in measuring parallax, and learn that stars are much farther away than anyone imagined.

In 1609, Galileo's telescope revealed a multitude of new discoveries, but Mars proved a difficult subject. The best of his telescopes magnified 33 times. It was not up to the task of showing any details on Mars.

The first telescopes gave narrow views, low magnification, and created distracting color fringes around bright objects. Fifty years after Galileo made his first telescope, Christiaan Huygens, a Dutch astronomer, overcame some of the problems. In 1659, he built exceptionally long telescopes that gave higher magnification and clearer views than Galileo's telescope. The unusual design had no tube. Instead, a long cord kept the eyepiece at the right distance from the lens, which was 123 feet in the air. It was so high people called it an aerial telescope.

With the long telescope, Huygens managed to make the first drawing of a feature on Mars. He saw a large V-shaped dark area, shaped like the subcontinent of India. He called it Syrtis Major, or "large bog." The feature is still a prominent one on Mars, although we know today it is not a bog.

If you look at photos of Mars, in some the V of Syrtis Major points up and in other photos it points down. Syrtis Major doesn't flip flop. Instead, some maps and photos of Mars (and other planets, too) have north at the top, while others have south at the top. Most telescopes used for astronomy give an upside-down view. Binoculars and terrestrial telescopes such as spotting scopes give right side up views. Turning the image right side up takes extra lenses that can reduce sharpness. Astronomers prefer a sharper view and leave the image upside-down. Some astronomy books and websites display photos that are upside-down so they will agree with what the astronomers see.

Very little was discovered about Mars until its close approach to earth in 1877. In the two hundred years since Huygens, telescopes had been improved to give sharper and more highly magnified views.

In 1877, Mars came within 35 million miles of earth, one of its closest approaches. Astronomers all over the world turned their telescopes to Mars to see what they could discover. However, except for one astronomer in the United States, no one looked for a satellite of Mars. They knew that Mars had no moon.

Despite careful study, no telescope user had ever seen a moon around Mars. In the 1860s, an astronomer in Europe explained why. He calculated where a moon, should it exist, would have to be located. A moon too far from Mars would be pulled away from the planet by the sun. A moon too close to Mars would be torn apart by tidal forces. Only within a certain band around Mars could a moon be found. Astronomers had already carefully inspected the area around Mars within that range. They found no moon.

The United States Naval Observatory in Washington, DC, had the largest telescope in the world at that time. It was a refractor with the main lens 26 inches in diameter. Its chief purpose was to gather information of interest to mariners (sailors),

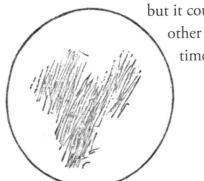

but it could be used for other observations as time allowed.

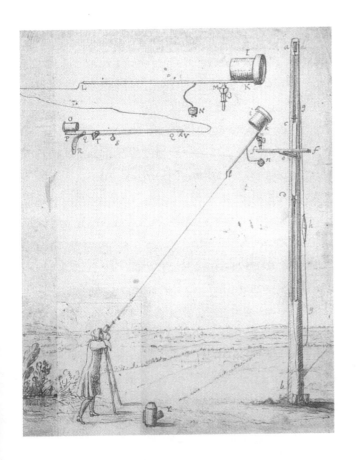

Christiaan Huygens using his tubeless aerial telescope; inset, his drawing of Syrtis Major

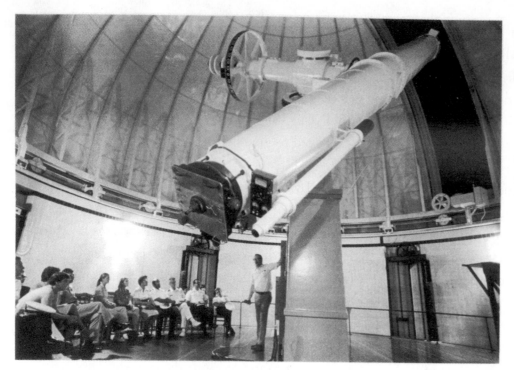

The telescope that Asaph Hall used to discover the moons of Mars, Phobos and Deimos; his wife, a skilled mathematician, assisted him in his calculations and discovery of the moons. The telescope is still in use and shown on public tours of the U.S. Naval Observatory.

Asaph Hall worked at the Naval Observatory. Hall came from a poor family and was largely self-taught. For a short time he did attend McGrawville College, a rural college in upstate New York. There he met Angeline Stickney, who taught mathematics. They married, and she encouraged her husband to pursue a career in astronomy.

In 1876, Asaph Hall used the Naval Observatory telescope to measure the rotation of Saturn. The planet had a great white spot, similar to the Great Red Spot of Jupiter. By timing the motion of the white spot, Hall calculated the rotation period of Saturn. He found it to differ by 15 minutes from the value stated in leading textbooks. Hall was appalled that astronomers had not noticed such a large error. Shortly after this, he came across the statement, "Mars has no moon." He began to doubt that statement, too.

In 1877, during the close approach to earth, Hall decided to use the powerful Naval telescope to search for a moon of Mars. He began the search early in August. The moon had to be small and dim, otherwise, it would already have been discovered. Hall began at the maximum possible distance for a moon and worked toward the planet.

He saw nothing that could be a moon. His last sweep took him so close to the bright planet that its glare had become a major interference. He closed the observatory, went home, and told his wife no moon could possibly exist.

While he had searched, his wife had redone the calculations. She was an especially skillful mathematician. Her calculations showed that a moon could orbit much closer to the planet than had been previously believed. Angeline Hall said, "Try it just one more night."

With her encouragement, he tried again. The next night, August 11, 1877, the seeing was horrible. The image of the planet bobbled about in air currents. Well after midnight, the seeing improved. Searching near the planet presented its own problems. Hall kept the bright planet just outside the field of view to overcome the glare. The Naval Observatory was along the Potomac River in a location known as Foggy Bottom. Hall was in a race against developing fog. At 2:30 a.m., he found a suspicious object. He barely had time to mark its position before the fog rolled in and obscured the view.

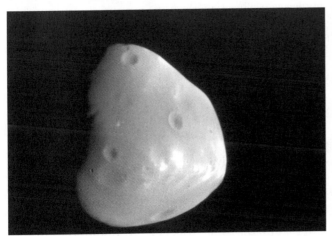

The two moons of Mars – Deimos (above) and Phobos (at right)

Good weather did not return until August 16. He spotted the object again. It traveled with Mars, so it was a satellite of that planet. The next night, while he waited for it to appear, he noticed another object close to the planet. It, too, proved to be a natural satellite of Mars.

In the space of a week, Asaph Hall had found two satellites of Mars. He named them Phobos (fear), and Deimos (panic). These are suitable names to accompany Mars, the god of war.

Jonathan Swift, an Irish-English writer, made an unusual prediction about the Martian moons in his book *Gulliver's Travels.* The book, published in 1726 and 150 years *before* Asaph Hall discovered the moons, tells the extraordinary adventures of the fictional character Lemuel Gulliver. In one of his travels, Gulliver visits the flying island of Laputa, inhabited entirely by scientists. He learns that "the Laputians have likewise discovered two lesser stars, or satellites, which revolve about Mars; whereof the innermost is distant from the center of the primary planet exactly three of his diameters, and the outermost five; the former revolves in the space of ten hours, and the latter in twenty-one and a half."

Jonathan Swift's description of the two moons made 150 years before their discovery turned out to be startlingly accurate. According to the fictional Laputian scientists, the closer moon was at a distance three times the diameter of Mars. Phobos is actually 1.4 times the diameter of Mars from its center. They said it traveled around Mars in 10 hours; its actual period is 7.7 hours.

Swift's Laputians did as well with the outer satellite. They said that it orbited at five times the diameter of Mars from its center. Deimos actually orbits at 3.5 times that diameter. They had it orbiting the planet in 21.5 hours. Deimos actually orbits in 30.3 hours.

Swift was as close as the usual experimental error in a new discovery. Was it just a coincidence, or did Swift know about the moons before the rest of the world? Actually, careful reasoning probably accounts for his unexpectedly accurate prediction.

Mars orbits between earth and Jupiter. Earth had one large moon, and Jupiter had four large moons. It seemed reasonable to Swift that Mars would have two moons. The telescopes of his day would have revealed larger moons, so he predicted exceptionally small ones hidden close to the planet. Swift took the time to figure their orbital periods, which gave his predictions a scientific basis.

Phobos is oblong in shape with dimensions of 17 miles by 12 miles. *Viking 1 Orbiter*

Percival Lowell is in the observer's chair of the 24-inch Alvan Clark & Sons refracting telescope. Lowell established the Lowell Observatory, and the telescope was installed in 1896.

dimensions of about 10 miles by 6 miles. It has such a weak gravity that a person could achieve escape velocity by sprinting and jumping off the surface. It has a dull gray surface that reflects only 7 percent of the light that strikes it. That makes it one of the darkest objects in the solar system.

Because of their near distance to Mars, Phobos and Deimos appear as unusual objects in its sky. Like our moon, they travel around Mars in a counterclockwise direction. Our moon travels so slowly that the rotation of the earth causes it to rise in the east and set in the west. But to an observer on Mars, speedy Phobos rises in the west, sets in the east, and rises again in just 11 hours on the same day. It skims around the equator of Mars at a height of only 3,700 miles above the surface. The orbit is so close to Mars that from the poles it cannot be seen. It is below the horizon all the time.

Deimos is farther away and travels more slowly at almost the same speed as the rotation of the planet. An observer on Mars would see the moon rise in the east and make a slow trip across the sky.

photographed Phobos in 1976. The photo showed its surface pockmarked by a large crater two-thirds the size of the moon itself. A large space rock struck it. A widespread system of fractures shows that the impact nearly shattered the moon. The crater was named Stickney to honor Asaph Hall's wife, Eangeline Stickney. After Hall discovered the satellites, she applied her mathematical skills in calculating their orbits.

Deimos is the outer moon and the smallest moon in the solar system. It is potato shaped, with

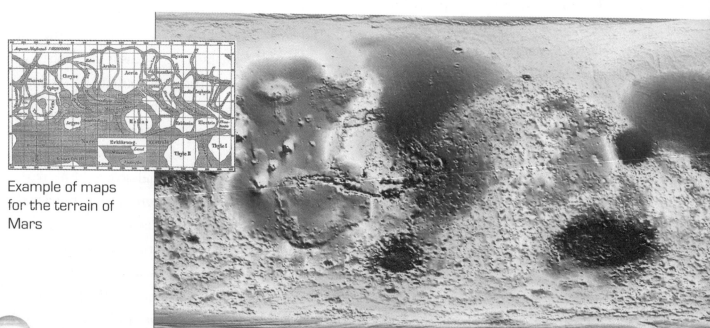

Example of maps for the terrain of Mars

Explorations of Mars include NASA's Exploration Rover Spirit, which captured images of the surface (below left); windstreaks in this image (below middle) help scientists calculate the wind's direction during the Mars Odyssey mission; Mars Global Survey image of a crater within the sandblown Martian dunes – notice the lack of dunes downwind of the crater that protects the area (below right).

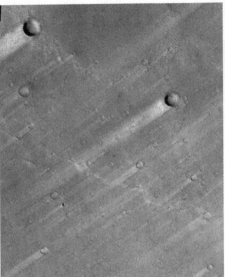
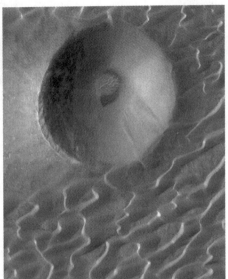

Despite its 30-hour orbit, it takes 65 hours to set in the west as it slowly falls behind the rotation of Mars. Because of its small size, it looks more like a bright star than a moon.

During the same year that Hall found the moons, an Italian astronomer, Giovanni Schiaparelli looked at the surface of Mars. His goal was to make a map of its surface. His telescope could see features no smaller than 200 miles across. Nevertheless, he marked narrow, dark lines on his map. He called them *canali*, the Italian word for channels, and gave them names of rivers on earth.

The word canali was mistranslated into English as canal. In English, a channel is a natural waterway, but a canal is an artificial waterway. Someone has to dig it. Most astronomers could not see the channels, but Schiaparelli's discovery gained the attention of newspaper reporters and writers of popular fiction. Reporters wrote stories claiming that intelligent beings lived on Mars. To survive on their desert-like planet, they had built a worldwide system of irrigation canals. Many people, including some astronomers, took the stories seriously.

A young American businessman named Percival Lowell read the reports. He had graduated with honors from Harvard. He was wealthy enough to pursue whatever interested him. He decided to build an observatory and dedicate it to the study of the Red Planet. He found a site with air clear and still near Flagstaff, Arizona. He called the location Mars Hill.

Even before the dome was finished to cover the telescope, Lowell was in the chilly air peering at Mars. Yes, he could see the long straight lines. He charted more than four hundred canals, and marked where the lines met as oases. He wrote several books about Mars. He argued that Mars changed color with the seasons because the Martians were growing crops.

Because detail was so difficult to see, astronomers could only speculate as to the meaning of what they saw — and often did. Their guesswork fueled speculation about Mars. Before long, many people became convinced that Martians existed. In 1898, H.G. Wells published *The War of the Worlds*. The book portrayed the Martians as invaders of earth who were trying to escape from their dying planet.

No earth-based telescope could settle the matter, so the idea of Mars as an abode of life

Technological advancements have enabled Mars missions to capture very detailed imagery of the planet's surface. This image revealed volcanic rocks, evidence of sulfur-rich minerals, and rocks that scientists think may be meterorites.

survived in the minds of some people until the 1960s. In 1965, a United States space probe passed by the planet. The photographs it sent back of the surface showed no canals. In 1971, *Mariner 9* took up an orbit around Mars. It photographed the entire surface of the planet. Scientists saw many spectacular features — but no canals.

The dark, narrow lines that astronomers had marked on Martian maps as channels did not exist. In some cases, several smaller features blurred together to appear as longer ones. Generally, however, nothing existed where channels had been seen. Trained observers had been fooled by optical illusions or wishful thinking.

In the 50 years after the first Mars probe, about 50 unmanned spacecraft were launched toward Mars, although a distressing number failed to reach their goal. But 21 made it all the way. Some took up orbits around Mars to photograph its surface. Others landed on Mars to test the composition of its soil. In 2012, no fewer than five Mars craft were actively seeking information about Mars — three orbiters and two remotely controlled vehicles on the surface. The answers to century-old questions started to come in.

Mars had vivid white caps at the poles. They appeared to change shape with the seasons, growing bigger in winter and smaller in

summer. Were the polar caps made of water ice? During spring, the color of Mars took on a deeper red. Features became blurred as if covered by blowing clouds of red dust. What caused the changing color, and did Mars have an atmosphere?

Mars is an interesting and exciting world, but vastly different from what astronomers originally imagined. We will look at the surface of Mars in more detail in the next chapter.

Discover

1. **For higher magnification and clearer images**

2. **Mars made a close approach to earth**

3. **They were small and close to the planet**

_____ 1. The Greek words *aster planetes* mean _____ stars.

A B C D 2. The five planets visible to the unaided eyes are Mercury, Venus, Mars, Jupiter, and (A. Neptune B. Pluto C. Saturn D. Uranus).

T F 3. To ancient astronomers, two puzzling aspects of Mars were that it changed in brightness and sometimes appeared to move backward.

A B C D 4. The astronomer who developed the sun-centered planetary system was (A. Copernicus B. Galileo C. Newton D. Ptolemy).

T F 5. Parallax of stars was easily visible in Galileo's telescope.

A B 6. The first feature on Mars seen with a telescope was named (A. Ursa Minor B. Syrtis Major).

A B 7. At first, the moons of Mars escaped detection because they were too _____ from the planet. (A. close B. far)

_____ 8. Of Mars and Earth, the one farthest from the sun is _____.

T F 9. Phobos means "far away" and Deimos means "delight."

A B 10. Compared to the other satellites in the solar system, the two moons of Mars are exceptionally (A. large B. small.)

A B C D 11. Phobos, the larger moon of Mars, has a crater named Stickney to honor Asaph Hall's (A. daughter B. father C. mother D. wife.)

A B 12. The person who built an observatory devoted to studying Mars was (A. Percival Lowell B. Giovanni Schiaparelli.)

T F 13. Photos taken by *Mariner 9* in 1971 showed that Mars was covered by a vast network of artificial waterways.

A B C D 14. The person who wrote about the two moons of Mars before they were discovered was (A. Christiaan Huygens B. Percival Lowell C. Jonathan Swift D. H.G. Wells).

EXPLORE MORE

Research the life of Johannes Kepler and write a two-page biography that summarizes his astronomical achievements.

The gravity on Mars is about 0.38 earth normal, and its year is 687 days. Calculate your weight on Mars. How many Martian years old are you?

How do you think the lower gravity of Mars would change a sport such as football or basketball? Imagine the game being played on Mars in a domed stadium with the atmosphere and temperature the same as on earth. Would the rules need to be adjusted? Would the playing field need to be modified? How do you think the lower gravity of Mars would change Olympic records in the shot put and high jump? Use current values for those records and the gravity on Mars to calculate what you think would be the Martian records for the same events. Would the time for the 100-meter dash be different on Mars?

Terrestrial Planets

Even before spacecraft landed on Mars and Venus and made photo fly-bys of Mercury, astronomers described those planets as being earth-like. They lumped the four planets nearest the sun into a category called terrestrial planets. The word terrestrial is from a Latin word *terra*, meaning earth.

The four inner planets — Mercury, Venus, Earth, and Mars — are alike in many ways. They greatly differ from the much larger outer planets — Jupiter, Saturn, Uranus, and Neptune. Astronomers call the outer four planets gas giants. They are huge. Jupiter, the largest of the gas giants, is 11 times the diameter of earth, the largest of the terrestrial planets. The gas giants have heavy atmospheres thousands of miles deep. The terrestrial planets, however, have rocky surfaces and a metallic core of mostly iron.

In the telescope, Mars appears to be earth-like, especially the light and dark markings on its solid surface and white polar caps. Mars

Explore

1. Is water found on Mars?

2. Which planet can be seen in the daytime with the unaided eyes?

3. Why did ancient people match Mercury with quicksilver?

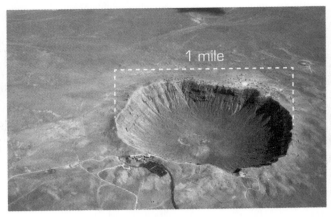
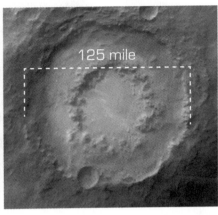

The Martian surface has revealed very large impact craters that are well-preserved. The Barringer Meteor Crater in Arizona on the left as compared to the Lowell Crater on Mars.

landers showed even greater similarities. Mars has valleys, mountains, canyons, sand dunes, craters, and volcanoes.

The impact craters on Mars are better preserved than those on earth and much larger. Heavy weathering has caused most impact craters on earth to all but disappear. The best-preserved one on earth is the Barringer meteor crater near Winslow, Arizona. Years ago a meteor about 150 feet across slammed into the desert and created a hole about one mile across and more than 550 feet deep.

Mars has thousands of craters larger than the one in Arizona. It has so many that only a few have been named. One named crater is Lowell Crater, in honor of Percival Lowell. Found in the southern hemisphere of Mars in the Aonia Terra region, Lowell Crater is about 125 miles across.

On earth, one of the best-formed and largest canyons is Grand Canyon. It is 277 miles long, 18 miles wide at its widest, and in places 7,800 feet deep. It extends about 1/100 of the way around the earth. But on Mars, Valles Marineris (named after the *Mariner 9* Mars orbiter that discovered it)

runs nearly one-fifth of the way around the planet. It is 2,500 miles long, 120 miles wide, and up to 23,000 feet deep.

How about mountains? Earth is almost twice the size of Mars. Earth's largest mountain is Mount Everest in the high Himalayas. It rises over 29,000 feet above sea level. Everest is dwarfed by the

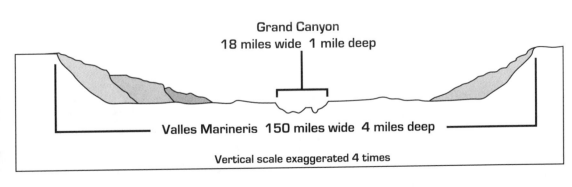

Grand Canyon
18 miles wide   1 mile deep

Valles Marineris   150 miles wide   4 miles deep

Vertical scale exaggerated 4 times

An immense "gash" can be seen in images of Mars. This is the Valles Marineris – see its size in comparison to the Earth's Grand Canyon.

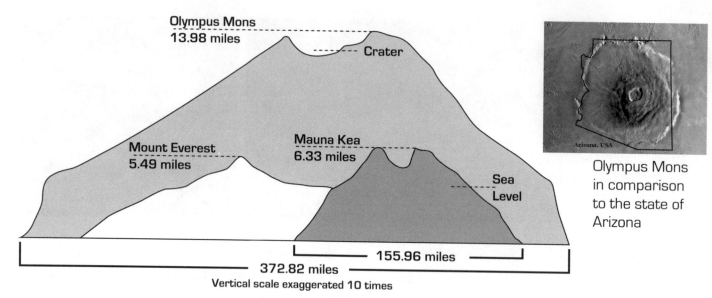

Olympus Mons
13.98 miles

Crater

Mount Everest
5.49 miles

Mauna Kea
6.33 miles

Sea
Level

155.96 miles

372.82 miles

Vertical scale exaggerated 10 times

Olympus Mons in comparison to the state of Arizona

A comparison of Mar's Olympus Mons to Earth's Mount Everest and Mauna Kea (if measured from its base deep below the ocean).

highest mountain on Mars. *Olympus Mons* (Latin for Mount Olympus) stands 78,000 feet above the surface of Mars. Astronomers believe it is the highest mountain on any planet. It is also a volcano, making it the largest volcano, too.

Earth does exceed Mars in several areas. Earth has more water, larger ice caps, and clouds. Earth is the only planet in the solar system that can have water in all three states — solid, liquid, and gas.

Mars does have water ice at its poles and in its soil but not liquid water. In certain areas, the temperature on Mars can rise above the melting temperature of water. However, even at that temperature, water on Mars cannot be liquid. Mars' thin atmosphere gives it a weak air pressure. Without the pressure to force water molecules together, they escape directly to a gas as the temperature rises.

The surface of Mars has harsh conditions; even so, it is still more earth-like than any of the other planets.

Record low and high temperatures on the two planets are closer than on any other world. The coldest spot on earth is Antarctica in winter. In July 1983, a Soviet research station about 800 miles from the geographic South Pole measured a record low temperature of −127°F. (Winter in the southern hemisphere occurs in July.) The average low at the poles on Mars is about −225°F. The hottest place on earth is Death Valley. On July 10, 1913, it reached a record high temperature of 134° F. Although Mars has not been fully explored, in some lowlands near the equator, scientists believe the temperature during the day may rise to at least 60°F, or maybe higher.

The temperature range on Earth is 261°F (-127°F to 134°F). On Mars the range is only 20 degrees greater, 285°F (-225° F to 60°F.)

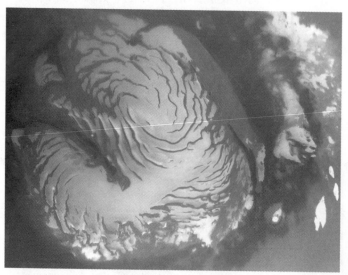

A polar ice cap on Mars

Earth's atmosphere has about 78 percent nitrogen, 21 percent oxygen, with traces of argon and carbon dioxide. Water vapor averages about 1 percent, and can be higher closer to the ground. The Martian atmosphere is about 95 percent carbon dioxide, CO_2, 3 percent nitrogen, and small trace amounts of water and oxygen.

Earth's atmosphere is about one-fifth oxygen. Oxygen is very active chemically and combines with most other elements. People often say that oxygen is necessary for life. But the reverse is also true. Without plants, oxygen would disappear from earth's atmosphere as it reacts with elements in the soil, especially iron. Mars' atmosphere is nearly devoid of oxygen. There are no plants restoring the oxygen.

Water is found in abundance on earth. If earth were a perfectly smooth globe and all the water from oceans, ice caps, glaciers, rivers, lakes, and all sources were spread evenly across its surface, the water would be about two miles deep. Mars has water, too, although not in as much abundance. Remote sensors show the white Martian polar regions contain water, as does the soil. Some estimates say that if all the water were spread across a perfectly smooth Mars, it would have a depth of about 30 feet.

The poles of Mars do contain an abundance of water. However, it is almost permanently frozen in place. The poles of each cap increase in size in the winter of each hemisphere. They grow not because of snowfalls but because carbon dioxide ice forms. Carbon dioxide freezes out of the air and adds to the size of the polar caps.

As summer approaches, carbon dioxide changes from the solid to a gas. A large volume of carbon dioxide gas creates vast winds that sweep down from the pole. Windstorms pick up dust and carry it across the face of Mars. The dust is fine as flour.

A helpful comparison chart on how our planet measures up to others in our solar system

Terrestrial Planets

	Mercury	Venus	Earth	Mars
Average distance from sun (AU)	0.39 AU	0.72 AU	1.00 AU	1.52 AU
Eccentricity of orbit	0.21	0.007	0.017	0.093
Orbital period	88 days	225 days	365.25 days	686.97 days
Rotation period	58.65 days	243 days	23 hr 56 min	24 hr 37 min
Diameter in miles	3,032 miles	7,522 miles	7,926 miles	4,221 miles
Diameter compared to earth	0.38/1.00	0.95/1.00	1.00/1.00	0.53/1.00
Tilt of axis	0°	177°	23.5°	25.19°
Mass compared to earth	0.055/1.00	0.82/1.00	1.00/1.00	0.11/1.00
Gravity compared to earth	0.38/1.00	0.90/1.00	1.00/1.00	0.38/1.00
Density in g/cm³	5.43 g/cm³	5.24 g/cm³	5.52 g/cm³	3.95 g/cm³
Albedo	0.06	0.75	0.37	0.15
Magnetic field	yes	none	yes	none

The eccentricity of an ellipse departs from a circle by the amount of separation of its two focus points [below]; the amount of eccentricity for Mars, Earth, and Mercury (right)

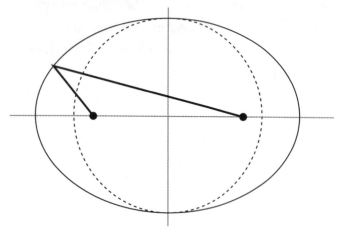

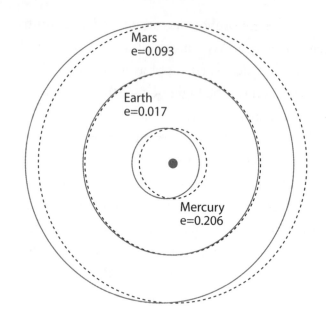

It contains a lot of iron oxide, or rust. The red rust gives the name "Red Planet" to Mars.

The table "Terrestrial Planets" compares Mercury, Venus, and Mars with earth. The first entry is AU, an abbreviation that stands for astronomical unit. It compares how far a planet is from the sun based on earth's distance from the sun.

The eccentric of an orbit is a measure of how much its orbit differs from being a perfect circle. An ellipse looks like a stretched-out circle. In a circle, the distance from the center to any point on the circle and then back to the center is always the same. Rather than a single center, an ellipse has two points called foci (singular focus). In an ellipse the distance from one focus to any point on the ellipse and then back to the other focus is always the same distance.

Eccentricity is how far an ellipse has been stretched out. If both foci are in the same place, then the ellipse is a circle and its eccentricity is zero, 0.0. If however it has been stretched so the two foci are far apart, then the eccentricity will be greater than zero, but always less than one, 1.0.

Most planets have an eccentricity near zero. But some comets have orbits that extend from near the sun to far beyond the orbit of Pluto. Their orbits have an eccentricity of closer to one. In all cases, the sun is at one of the foci. The other focus is empty.

Because of a planet's eccentricity, a planet will be closer to the sun during part of its orbit. The closest approach to the sun is called the perihelion. Helion is from *helios*, a Greek word meaning the sun. (The moon's orbit is eccentric, too, but because it orbits the earth, the moon's close approach to the earth is called its perigee. *Gee* is a root word meaning earth, as in geology.) The greatest distance a planet gets from the sun is called the aphelion.

Mars' orbit is more eccentric than the earth's orbit, 0.093 versus 0.017.

The orbital period of a planet is the time to revolve around the sun, while the rotation period is the time to rotate once on its axis. Because a planet is moving, the rotation period and the length of a day can be different. After the sun comes up, the planet moves, and the planet has to rotate a few minutes longer to put the sun back in the same location a day later. For earth, it takes about four minutes for the sun to catch up. Earth's day is four minutes longer than its rotation period. For Mars, Mars' day is two and a half minutes longer than its rotation period.

The diameter of Mars is only about one-half the diameter of earth (4,221 miles compared to 7,926 miles). The tilts of their axes are about the same (25.19° compared to 23.5°). Its mass is only about one-ninth of earth's mass (0.11 compared to 1.00). Gravity on Mars is a little more than one-third of earth normal (0.38 compared to 1.00). Mars is less dense than earth (3.95 g/cm³ compared to 5.52 g/cm³), and only about half as reflective (0.15 albedo compared to 0.37 albedo). Mars, like the moon, has no magnetic field. A compass would not work on Mars.

The Red Planet is a colorful contrast to Venus, our nearest neighbor sunward. Venus is always a vibrant, sparkling white.

On a clear evening after the sun sets, while the sky still glows, what appears to be a single bright star sometimes appears in the west. It comes out before any other stars. After the sky darkens it shines with a brilliant light far brighter than the brightest star. Under really dark skies without streetlights or the moon to interfere, the "star" is so bright it can cast a shadow.

The light is not a star. Instead it is Venus, the brightest object in the sky other than the sun and moon.

Venus first appears in the evening sky very low in the west shortly after the setting of the sun. From night to night it appears higher in the sky after sunset. It climbs until it is halfway overhead. It shines in the dark sky for three hours before it sets. As the nights pass, Venus draws closer to the sun. Finally, it sets at the same time as the sun and disappears from sight.

A determined amateur stargazer can awaken early and look for it in the morning sky. At first it rises just before sunrise. As days pass it comes up well before sunrise until it is once again halfway to the zenith (the portion of the sky directly overhead). Once again it dips back toward the sun, until it rises at the same time as the sun and becomes lost in its glare.

Because Venus could only be seen early in the morning or early in the evening, it became known as the Morning and Evening Star. At first, people believed the Morning and Evening Star were two different bodies. But the Greeks realized they were the same object. They identified it as a planet and gave it the name Aphrodite, the goddess of love. Venus is the Roman goddess of love. Venus is the only major planet with a feminine name.

Venus is so bright it can be seen in the daytime with the unaided eyes. The best way to find it in the daytime sky is to watch it as the sun comes up. As the sun rises, the stars fade away. But Venus will still be visible as a pinpoint of bright light, even with the sun above the horizon. It helps if Venus is near the moon because that helps your eyes focus at the far distance. Otherwise, in a clear blue sky, they will focus closer and Venus will blur out. Later in the day, first find the moon, and then look near it for Venus.

Other than tracking Venus and plotting its orbit, early astronomers could learn nothing more about Venus until the invention of the telescope.

The moon and Venus are among the most easy to identify celestial objects and can be readily seen from the Earth's surface.

When Galileo turned his telescope to Venus, he saw it had phases like the moon. The phases change as Venus orbits the sun, from a thin crescent to a full phase. The full phase occurs when Venus is opposite the sun from earth and appears the smallest. During its crescent phase, its apparent size from earth is at its largest and the planet is brightest. It is a remarkable sight in a telescope.

It is best viewed in the crescent phase and appears to be about 1/30th the size of the moon. Venus is, of course, more than three times larger than the moon, but because of the distance it seems far smaller. Some sharp-eyed observers claim to have seen the crescent horns of Venus with their eyes alone. Others dismiss such claims. But the slightest magnification brings them out. Even a toy telescope that gives 2x or 3x magnification is enough. A larger telescope and higher magnification makes the planet look larger but reveals nothing more than the phases. A thick layer of clouds always hides its surface.

Despite its hidden face, Venus remained a fascinating object to astronomers, especially when they calculated its orbit and found that it would occasionally pass across in front of the sun. Venus' orbital plane is set at a small angle to earth's. They are inclined to one another by about 3.4 degrees. Most of the time, Venus passes above or below the sun. On rare occasions, however, Venus passes across the face of the sun, an event called a transit of Venus. The transits only take place about twice every 100 years.

The first person to see a transit of Venus was 20-year-old Jeremiah Horrocks, who lived in England. He served as a full time clergyman's assistant but devoted his spare time to astronomy. Astronomers calculated that Venus would make a near miss of the sun sometime in 1639. In his spare time Horrocks redid the calculations and became convinced that Venus would actually transit the sun sometime on November 24, 1639.

Clouds and his other duties interfered looking for it until later in the afternoon. He ran to his room where he had built a device to watch the event. He wisely chose not to look at the sun directly. He projected the image from a telescope on a sheet of paper. At first he saw no evidence of the planet and thought he had missed it.

A closer look at Venus – the planet is often hidden behind a large veil of clouds as shown (above left); radar imagery was combined to create this image (above right) of Venus' surface beneath the clouds.

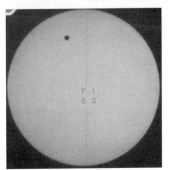

Illustration of Jeremiah Horrocks observing the first recorded transit of Venus across the Sun in 1882; at left below is his drawing of the transit; at right, the 2004 transit of Venus

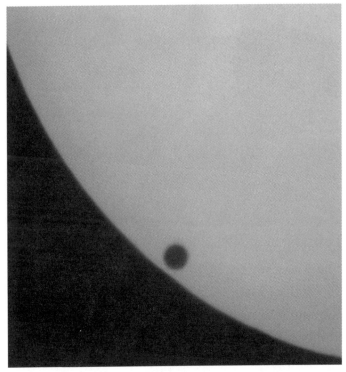

Finally, at 3:15 p.m., he saw the black dot of Venus moving across the face of the sun. It covered only 1/1,000 of the sun's face. He watched and took measurements until the sun set.

At that time, astronomers gave solar system distances in astronomical units (AUs). They had developed a scale model of the solar system and could calculate the relative location of a planet compared to earth's distance. They needed to learn the distances in actual values such as millions of miles.

Horrocks believed Venus would provide a solution. He proposed to measure parallax of Venus against the sun as seen from several different locations on earth. Transits come in pairs, eight years apart, with a long gap between the pairs. The next pair of transits occurred in 1761 and 1769; followed by transits in 1874, 1882, 2004 and 2012. No more will occur until 2117 and 2125.

Horrocks would not be around to see the next transit. It would not occur for another 122 years. But astronomers got ready as 1761 approached. They loaded their telescopes and traveled to Siberia in Russia, Newfoundland in North America, the Cape of Good Hope at the tip of Africa, and Tahiti, a lonely island in the Pacific Ocean. They planned to accurately measure their location in latitude and longitude and exactly time the start of the transit.

Timing first contact of Venus with the face of the sun proved difficult. Venus' thick atmosphere gave a glowing ring around the planet. Despite careful planning, the results could not be fully trusted. Another transit took place only eight years later in 1769, so they tried again. This time observations were made in 62 different locations. They narrowed the size of earth's orbit to somewhere between 97 and 93 million miles.

Once earth's distance was known, astronomers calculated the size of the orbits of other planets, including Venus. Venus is about 67 million miles from the sun. Its orbit takes it within 25 million miles of earth, closer than Mars by seven million miles. Venus, not Mars, is earth's nearest planetary neighbor.

Despite being so near, the surface of the nearest neighbor to earth remained a complete mystery to astronomers. Clouds hid its face.

Astronomers believed the clouds were water vapor. Some thought the water evaporated from a gigantic ocean that covered the entire surface of the planet. Others predicted that an unbroken marsh and swamp covered its surface. Rain fell a

Another example of imaging radar being used to get surface details of Venus. This image was captured by the Magellan spacecraft during its 1990-1994 orbital mission. Beneath the planet's sulphuric-acid clouds, the imaging radar mapped this three-dimensional view of the volcanoes Gula Mons (right) and Sif Mons (left).

constant torrent. It became popular to think of Venus as a watery, warm, and humid earth.

In the 1950s, scientists learned how to measure the heat that radiated away from a planet. They looked at the infrared part of the spectrum. The visible spectrum consists of the colors red through violet. Infrared radiation (light waves longer than red and below red in the spectrum) is invisible to human eyes but can be felt as heat rays. American astronomers outfitted telescopes with infrared sensors and measured the temperature of the clouds on the night side of Venus. More heat escaped from the planet than they expected. The instruments revealed that the clouds were surprisingly hot, about 625°F.

Most astronomers dismissed this report. It did not agree with the currently accepted ideas about the planet.

A few years later, space probes proved the measurement was not a fluke. The original ideas about Venus were completely in error. Venus is hot, far hotter than any astronomer had imagined. The temperature of the surface of the planet averages about 890°F. It is hot everywhere — at the equator, at the poles, on the dayside and on the night side. It never cools down. Its temperature of 890°F is three times as hot as an oven roasting a Thanksgiving turkey.

Clouds of carbon dioxide and not water vapor swirled around the planet. The carbon dioxide clouds contained a dash of hydrogen sulfide,

a gas that smells like rotten eggs. The atmosphere is especially dense, as much as 90 times earth normal.

The view from the surface of Venus would be a gloomy one. Anybody on its surface would have to take the word of other travelers that beyond the clouds were a sun, planets, and stars. Venus has no moon, but even if it did, a surface dweller would not be able to see it.

Not only is the surface a gloomy one, it is difficult to land on and get photographs. In the 1960s, the United States and Soviet Union engaged in what became known as the Space Race. They sent unmanned vehicles to the moon, Mars, Venus, and elsewhere. In 1962, the United States craft *Mariner 2* made a fly-by of the planet.

In 1975, Soviet spacecraft *Venera 9* and *Venera 10* parachuted to the surface of Venus. They managed to send back photos for an hour before the heat and high pressure destroyed their instruments. The black and white images showed a surface covered with rocks one to three feet across. In 1982, *Venera 13* sent back color photos for two hours before it too succumbed to the difficult conditions.

Building craft to survive on the planet proved a difficult task. Ordinary glass in camera lenses would melt in the high heat. Instead, the lenses had to be made of quartz. Rather than send craft to die on the surface, the United States designed spacecraft to go into orbit around the planet. The orbiters dropped expendable probes into the atmosphere and used radar to peer below the clouds.

The United States *Magellan* spacecraft made a radar map of the surface.

The surface of Venus is rocky and much of its surface appears to be made of basalt — cooled lava. Like Mars and earth, it has volcanoes, lava flows, craters, and two highlands that would be continents if it had a watery surface.

The radar also showed Venus' rotation rate. Venus takes 243 days to rotate once, far longer than the rotation rate of Mars and earth. Yet it needs only 225 days to go around the sun. If the clouds did not obscure the sun, a person on the surface of Venus would see the sun rise in the west and set in the east, exactly opposite the way it moves on earth and the other planets.

Refer back to the table of terrestrial planets and see how Venus compares to earth. Notice that the tilt of the axis is 177 degrees, nearly upside-down. (A tilt of 180 degrees would be exactly upside-down.) Astronomers list the tilt this way to account for the backward rotation. To demonstrate this, make a fist with your right hand with the thumb pointed up. As seen from above the fingers curl in a counterclockwise direction. Now turn your fist upside-down. The fingers curl in a clockwise direction, the same direction as Venus' rotation. Astronomers think of Venus as an upside-down planet.

Mercury is the fourth terrestrial planet and the one closest to the sun. People of ancient times associated metals with the planets. That matched the sun with gold, the moon with silver, and Mars with rusty iron. For the planet Mercury, they matched it with what they called quicksilver — a heavy liquid metal that darted around when held in their hands. The modern name for quicksilver is mercury. (We also know today that it should not be held in the hands. It is a heavy metal and therefore poisonous.)

Like quicksilver, the planet Mercury moved rapidly in the sky. The Greeks thought of Mercury as a runner carrying important messages. They pictured him with wings on his feet. Their name for Mercury was Hermes.

Mercury orbits the sun at only 36 million miles. It travels faster than any other planet. Like a fast runner on an inside track, Mercury whirls around the sun in only 88 days.

Like Venus, Mercury is called a morning and evening star. But it does not share Venus' brilliance nor does it rise as high in the sky. Mercury never gets more than 28 degrees from the sun, only half as far as Venus. After sunset, it is low in the western sky. It must compete with twilight glow and is often hidden by mountains or tall trees. In the morning it is the same number of degrees above the horizon at sunrise.

Mercury is an elusive planet to see with the eyes alone. It is high enough to be seen for about a week at a time, and only for three or four times a year. Copernicus is reported to have never seen it. To see Mercury in the evening sky, look for a western horizon clear of obstructions. As usual, it is easier to find a planet when it is close to the moon, and Mercury is no different.

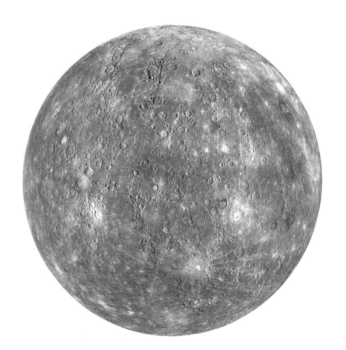

Features on the surface of Mercury include the crater at center near the bottom; this is Debussy – it is distinctive for the rays extending from it.

Studying the surface of Mercury from earth-based telescopes is difficult. It has phases like Venus, but that is about all that can be seen. In the late 1700s, William Herschel, who discovered the planet Uranus, detected some features on the planet. Every time he turned his telescope to Mercury, he saw the same features. He believed the planet always showed the same face to earth and the other face to the sun.

According to Herschel, Mercury had one face locked toward the sun in the same way that the moon has one face locked toward earth. If that were the case, then Mercury's day and year would be the same — 88 days. Over the next 100 years, other astronomers agreed with him.

They used his observations to build up an incorrect depiction of Mercury's surface. Having one side facing the relentless heat of the sun would make Mercury hotter than anywhere else in the solar system except the sun itself. The lighted side would be hot enough to melt lead. The other side would forever face the cold and dark of deep space. Heat would radiate away, leaving that side of Mercury one of the coldest places in the solar system. However, this picture of Mercury was incorrect.

Mariner 10, launched in 1973, went into orbit around the sun in such a way as to pass close to Mercury on three occasions. It took pictures of the planet and sent back measurements of its temperature. The spacecraft proved that the planet rotated on its axis in 58.7 days. During the time to go around the sun two times, Mercury spins on its axis three times. The rotation moderates the temperature. Mercury has only a very slight tilt to its axis. Its poles are always in shadow. Deep craters at the poles have water ice on their floors.

Venus turned out to be hotter than Mercury. Venus, and not Mercury, has a surface hot enough to melt lead.

Mercury's surface is heavily cratered. It looks more like the moon than any of the other planets. It even has bright rays going out from some of the craters as the moon does. Because of the dark material on its surface, it reflects light poorly. It has the lowest albedo of the terrestrial planets.

Refer to the table of terrestrial planets for more information about Mercury. As you look at the table, compare and contrast the planets. Which has the highest gravity; which has the lowest? On which planet would you weigh about the same as you do on earth?

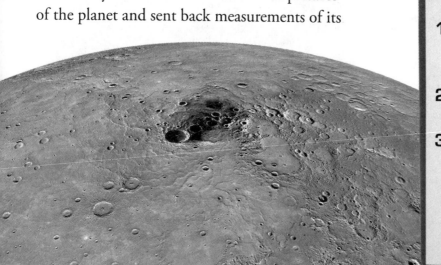

Excellent image of Mercury's North Pole

Discover

1. Mars has water as ice at its poles

2. Venus

3. Mercury, like quicksilver, moves quickly

Questions

1. List the four terrestrial planets in order from the sun.

T F 2. Mars has craters larger than any on earth.

T F 3. The Grand Canyon on earth is the largest in the solar system.

A B C D 4. The most common gas in the atmosphere of Mars is _____.
(A. carbon dioxide B. nitrogen C. oxygen D. water vapor)

A B 5. The chemical compound that gives Mars its red color is (A. iron oxide B. sulfur).

A B 6. Mars travels around the sun in _____ orbit. (A. a circular B. an elliptical)

A B 7. The larger planet is (A. Earth B. Mars).

_____ 8. The brightest object in the sky other than the sun and moon is _____.

T F 9. Venus is the only major planet with a female name.

_____ 10. Galileo's telescope showed that Venus has _____ like the moon.

A B 11. Earth's nearest planetary neighbor is (A. Mars B. Venus).

A B C D 12. Spacecraft landing on Venus failed because of the planet's (A. acid rain
B. high temperature C. intense cold D. lack of sunlight to power solar cells).

A B C D 13. Mercury takes _____ days to orbit the sun. (A. 88 B. 225 C. 365 D. 687)

T F 14. At one time, astronomers incorrectly believed Mercury kept the same side facing the sun.

EXPLORE MORE

Research the life of Copernicus and write a two-page biography that summarizes his astronomical achievements.

The gravity on Venus is about 0.70 earth normal, and its year is 225 days. Calculate your weight on Venus. How many times has Venus gone around the sun since you were born?

Mars and Mercury both have impact and volcanic craters. How do impact and volcanic craters differ in shape? Draw each in profile (side view), and from above.

On which of the terrestrial planets (other than earth) do you think would be the easiest to build a permanent settlement? The hardest? Research some of the plans for making interplanetary voyages and permanent settlements on Mars.

Using an appropriate scale based on the size of the paper you have, draw circles to model the size of the terrestrial planets. Nest the circles inside one another with a common center. Label each circle to show the planet it represents. As a guide to diameter use the table of terrestrial planets. Color each circle based on the color that you think best represents that planet.

Jupiter

Jupiter is a grand planet, the largest in the solar system. Although more distant than Mars, Jupiter is usually brighter. Mars can edge it out slightly in brightness when closest to earth, but then Mars dims as it and earth separate. Jupiter, on the other hand, stays bright year around. It shines with a robust, unwavering light. It is easy to spot because even to the unaided eyes it gives the impression of a big planet.

The first observers of the night sky noticed that Jupiter moved with ponderous dignity. It took a year to travel from one constellation to the next. It commanded the night sky for months at a time. Mercury barely made it out of the sun's glare. Bright Venus never managed to climb directly overhead. Mars could shine directly overhead, but could not for long rival the steady, bright light of Jupiter. Early stargazers considered Jupiter the king of the planets. The Greeks named it "Zeus" after their chief god, and the Romans called it "Jupiter" after the chief god in their pagan religion.

Explore

1. **What does the word *satellite* have to do with court officials?**

2. **What storm has lasted 400 years on Jupiter?**

3. **Which moon of Jupiter has an oxygen atmosphere?**

Jupiter is the largest of all the planets in the solar system. By the 1600s, astronomers realized it had to be huge because of its brightness and distance from the sun. They calculated its distance as five times as far from the sun as earth. As light spreads out from the sun, it covers a surface that grows larger by the square of the distance. In other words, light grows weaker by the square of the distance. By the time light reached Jupiter it was 25 times (5 x 5 = 25) weaker than the light striking earth. The light then grew dimmer as it reflected from Jupiter and traveled back to earth.

For it to shine so brightly yet be so distant, it had to be a big planet. And it is. Jupiter has a diameter 11 times the diameter of earth.

But diameter alone is not the only way to measure size. Surface area goes up by the square of the diameter. A ball with twice the diameter of a smaller one has four times more surface area: 2 x 2 = 4. One three times greater in diameter has nine times the surface area: 3 x 3 = 9. Jupiter has 11 times the diameter of earth. Its surface area is 121 times that of earth: 11 x 11 = 121.

The disparity in size between Jupiter and earth is even greater based on volume, the amount of space inside an object. The volume of a globe increases by the cube of the diameter. A ball twice as large has a volume eight times as great: 2 x 2 x 2 = 8, while a ball three times as large has a volume 27 times as great: 3 x 3 x 3 = 27. Jupiter has 11 times the diameter of earth, so its volume is 1,331 times a much: 11 x 11 x 11 = 1,331.

Because of its large size, Jupiter is easy to observe in a small telescope. It presents a large disc and is visible all year around except for a month or two when it is behind the sun as seen from earth. Jupiter at its greatest distance from earth presents a globe larger than Mars at its closest.

When Galileo turned his telescope to Jupiter he found the most astonishing of all his discoveries. On January 7, 1610, Galileo examined the bright

Galileo's notes and drawings on the satellites of Jupiter

disk of Jupiter. His telescope made objects seem 30 times closer than with natural vision. It also gathered light, making dim objects appear brighter. Next to the planet, he spotted what he supposed at first were three fixed stars, totally invisible because of their smallness, all close to Jupiter, and lying on a straight line through it. He sketched what he saw:

＊ ＊ **O** ＊

When he viewed Jupiter the next night, he found three stars again, but this time they had shifted position. All three were on the same side:

＊ ＊ ＊ **O**

Night after night for more than a month he tracked the shifting position of the little "stars." He expected Jupiter to move from night to night, and it did. But he expected it to move away from the stars, and it didn't. Instead, Jupiter carried the little lights with it. One night he saw four of the bodies.

＊ ＊ ＊ **O** ＊

He watched for five hours, and he could actually see that the one closest to Jupiter had drawn closer still. He grasped the truth. These were not fixed stars, but four bodies belonging to Jupiter and going around it. He continued to observe Jupiter through February 1610. He recorded what he saw every night in his notebook. He published his discoveries in a book called *Message from the Stars*.

43

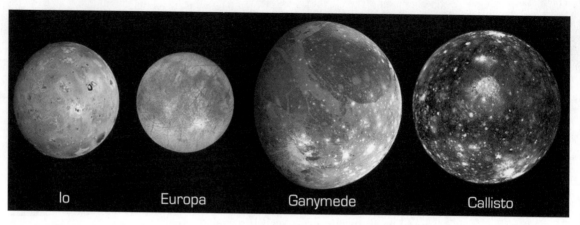

Jupiter's four largest moons and relative size.

Io Europa Ganymede Callisto

Johannes Kepler read *Message from the Stars*, Galileo's book about his many discoveries with the telescope. Kepler realized that bodies that go around planets needed a name. In his day, the minor court officials who attended to the wishes of their king were called satellites. Here was Jupiter, the king of the planets, who ruled over four moons. Kepler called the new objects satellites. Astronomers quickly agreed to that term, and it has become the accepted name for any object that orbits around another one.

Galileo knew his discovery of the moons would cause an uproar among astronomers who believed the earth-centered planetary system. They used the fact that the moon orbited the earth as proof the earth had to be unmoving at the center of the planetary system. They argued that the earth could not move around the sun. If it did, the earth would leave the moon behind. The moons of Jupiter were clear and convincing evidence otherwise. Jupiter moved, yet it did not run away from its moons. With the discovery of the moons of Jupiter, astronomers quickly abandoned the earth-centered planetary system in favor of the sun-centered one.

Binoculars are powerful enough to show the satellites of Jupiter. The two outer moons are easily visible. The other two can be glimpsed as well when favorably placed, although they can be lost in the glare of the planet. When all four are visible at once, they are found in a straight line going out from the equator of Jupiter.

Callisto is the outermost moon. It is 3,000 miles in diameter, about the same size as Mercury. The next one going in toward Jupiter is Ganymede. With a diameter of 3,270 miles, it is the largest satellite in the solar system. It is larger than the planet Mercury. However, it has only half its mass. Ganymede and the other moons are made of material less dense than the rocks and inner core of Mercury.

Second from Jupiter is Europa, the smallest of the four moons, but still large at 1,940 miles in diameter. The fourth moon, Io (eye-oh), orbits closest to Jupiter and is 2,263 miles in diameter, or a little larger than earth's moon. Io spins around Jupiter very quickly, making a complete orbit in about 42 hours. In the space of a night's observation, it moves from one side of Jupiter to the other.

Image showing Jupiter and three of its moon: Callisto and Ganymede to the left of Jupiter, and Europa on the right of Jupiter.

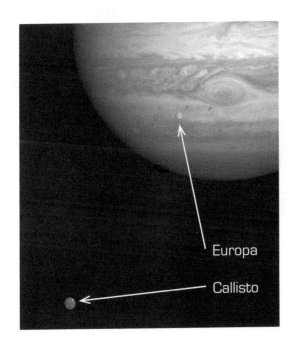

Europa

Callisto

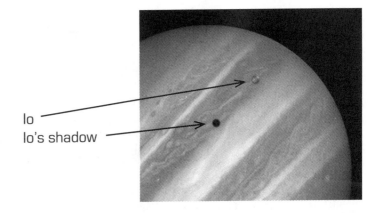

Io
Io's shadow

A small telescope makes watching the dance of the moons easier and more enjoyable. A small telescope at 40x will clearly show all four satellites, although at any one time a moon may be behind the planet, disappear from sight when directly in front of the planet, or wink out as it passes into the shadow of Jupiter. Other moons orbit Jupiter, more than 60 at last count, but they appear to be captured space debris. They are too dim to be seen in all but the largest telescopes, so they do not confuse the amateur telescope user by getting in the way. A "star" that travels with Jupiter and goes around it must be one of the four satellites that Galileo saw.

Even without the moons, Jupiter would be a rewarding object for small telescopes. At about 200x magnification all sorts of detail can be seen. Sometimes the shadow of one of the moons can be seen as it falls on the face of the planet. One of the great puzzles of Jupiter is also visible — the Great Red Spot.

The Great Red Spot was first observed in 1671 by Giovanni Cassini. He was an Italian who began his career at the University of Bologna in Italy. He made a multitude of discoveries. In 1666 he measured the rotation rate of Mars. In 1668 he published a table listing the orbits of Jupiter's four moons. His success in Italy came to the attention of King Louis XIV of France. The king invited Cassini to Paris in 1669. Cassini served as chief astronomer in Paris for 40 years.

He didn't mind tackling difficult and long-lasting projects. In the 1670s he set out to make a topographic map of France. A topographic map not only shows distances but also ridges and valleys of the terrain. After his death in 1712, his son,

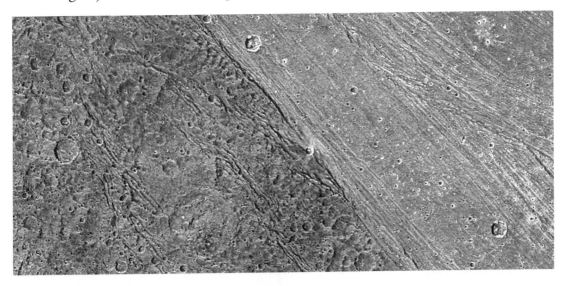

An image showing the terrain on Ganymede

Voyager 1's photo of the Great Red Spot, a storm on Jupiter that can be likened to the worst hurricanes on Earth. It's an ancient storm that has been observed for over 400 years. It is so large that three Earths could fit inside it.

and then his grandson, carried on the work. They finished the task in 1793 — more than 120 years after he began.

He and a fellow astronomer, who traveled to Africa, measured the parallax of Mars. In 1672, he calculated its distance from earth. From that he could estimate the dimensions of the solar system. He set the distance of earth from the sun at 87 million miles, off by six million miles, an error of 7 percent.

In 1671, Cassini observed a "persistent spot" on Jupiter. Robert Hooke, an English scientist, had also seen it at about the same time. This was the first mention of the Great Red Spot.

The telescope he used at the Paris Observatory had a main lens only 2.5 inches in diameter — no larger than a small amateur telescope today. Opticians had not yet learned how to avoid the prism effect of lenses. Large lenses produced a smear of rainbow-colored fringes that blurred the object being viewed.

The Great Red Spot and other features Cassini observed can be seen in a small telescope today, which is generally of higher quality than the lens he used. A small telescope shows broad reddish belts that run parallel to the equator and cover

Jupiter's visible disk. Lighter-colored regions separate them. The belts on either side of the equator are called the north and south equatorial belts. The Great Red Spot floats below the south equatorial belt in a lighter band. It does vary in color. In 1878 it turned a vivid brick red. That is when it gained the name *Great Red Spot*.

The Great Red Spot spins in a counterclockwise direction with the winds at its edges going at 250 miles per hour. Some astronomers call it a long-lived storm, like a hurricane on earth, but one that rotates in the opposite direction. Normally, when a storm over the ocean makes landfall, it weakens and comes apart. The fact that the Great Red Spot has raged for at least 400 years suggests that it never passes over a solid surface. Although Cassini first saw it in 1671, it could have existed years before then. No one knows when it began or how long it will last.

No one knows what causes the red color, either. On earth, red phosphorus (like on the tip of a match) and sulfur compounds have a red color. Perhaps the clouds of the spot contain phosphorus or sulfur.

The spot does vary in color, from almost brick red to pale salmon. Its top normally stands about five miles above the surrounding white clouds. Sometimes it appears to sink below the white band and nearly disappears.

When the color becomes less vivid, amateur astronomers resort to tricks to see it. Colored filters can make some colors more noticeable and weakens other colors. Viewing Jupiter through a light green or blue filter makes the spot stand out. Digital photos can also be enhanced to emphasize one color and diminish others. Even without the filters, the spot can still be detected because it causes an indention in the south equatorial band.

The Great Red Spot is south of Jupiter's equator. However, photos are often printed upside down to match the inverted view of a telescope. Because people are used to seeing maps with north at the top, they may think the spot is also in the northern hemisphere of Jupiter.

Cassini noticed that the Great Red Spot drifted across the face of Jupiter until it disappeared around the other side. Later, it reappeared. He timed its passage and used it to measure how quickly Jupiter rotates. It spins on its axis in just under 10 hours, faster than any other planet.

Jupiter rotates in less than 10 hours (9 hr., 55 min., 30 sec.), which is amazing considering its size. Jupiter is 11 times bigger than the earth, and so has a circumference 11 times greater. Yet it rotates on its axis in less than half the 24 hours earth takes. Jupiter's equatorial surface moves far more rapidly than earth's.

Jupiter's fast rotation causes it to bulge at the equator. The faster a planet rotates, the greater the force causing it to expand at its equator. The earth rotates once in 24 hours, and it is about 25,000 miles around at the equator. An object on the equator is traveling at about 1,000 miles per hour. The earth is slightly flattened due to this rotation. At the equator, a person is 26 miles farther from the center of the earth than at the poles. Because of its

fast rotation, Jupiter has an even greater bulge at its equator.

The speed at the equator of Jupiter can be found by dividing the circumference of Jupiter by its rotation rate. Speed equals distance divided by time. Dividing its circumference of 279,000 miles by its rotation rate of 10 hours gives a velocity at the equator of 27,900 miles per hour. Material along the equator is thrown outward. An object at the equator is 3,000 miles more distant from the center of the planet than at the poles.

Cassini also found other features in the cloudy bands around Jupiter, but their rotation rates varied depending on their location. Instead, the equator rotates more quickly than areas north or south of the equator. He concluded that Jupiter does not rotate as a solid body.

Cassini also published times of eclipses of Jupiter's satellites as a possible aid to ship's navigators. To locate a ship on the ocean far from land, a navigator must calculate its latitude and longitude.

Latitude — distance north or south of the equator — could be accurately figured in a variety of ways. For instance, the distance of the North Star, Polaris, above the equator gave the latitude. If Polaris was 14° above the horizon, then the ship was at 14° north latitude, or about 966 miles north of the equator. Polaris is not quite at the North Pole, but the navigator could make a correction for the slight error. The sun could also be used, but doing so involved more calculations. It had to be observed exactly when at its highest, and corrections had to be applied based on the time of year. By the late 1600s, a navigator could measure the ship's latitude within 1/6th of a degree, about 12 miles.

Longitude — the distance east or west — could only be measured by comparing local time as given by the sun at noon with a clock set to the time at the homeport. For British sailors that was Greenwich, England, near the Thames River.

Sextants and how they work

The sextant is a navigation instrument used primarily to find latitude, although it can be used for other purposes, too. The word sextant means one-sixth of a circle. Until the 1400s, European ships sailed within the confines of the Mediterranean Sea. Those that ventured out into the wide ocean sailed along the shore within sight of land.

However, in the 1450s, ships began sailing farther from shore. To locate their position, they needed to know their distance north or south of the equator. At night, Polaris guided them. The North Star's position above the horizon was equal to their position above the equator.

During the day, they used the sun. Its position gave them the same information as Polaris except the distance in degrees had to be subtracted from 90 degrees. If the sun were directly overhead, 90 degrees, then they were exactly at the equator, 0 degrees latitude. This method was only exact on the first day of spring and the first day of fall at noon when the sun was directly over the equator. On all other days they had to calculate corrections based on time of day and time of year.

While looking through the sextant, the ship's navigator would move an arm that used mirrors to lower the image of the sun so it appeared to be on the horizon. (Dark filters prevented damage to the eyes.) The process was called shooting the sun. A scale showed how far the arm moved and from that the latitude could be calculated.

The sextant came into wide use by seafarers, as well as other people. David Livingstone used one as he explored the continent of Africa. When he did not have a clear level horizon, such as one found on the ocean, he would fill a large, flat pan with water. He used the level of the water as a horizon.

Even as late as the 1960s, large airplanes had a clear dome at the top of the plane. The navigator would stand with his head and shoulders in the dome and use the sextant to measure sun or star positions to find the exact location of the airplane.

A sextant does not require electricity. Ships still carry one in case electronic equipment fails.

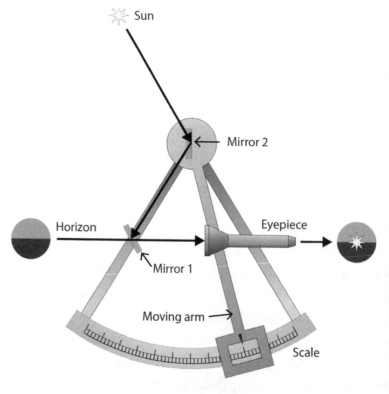

Parts of a sextant used in "shooting the sun"

Ole Roemer's studies (above) on the variation of moon eclipses helped him discover he could calculate the speed of light

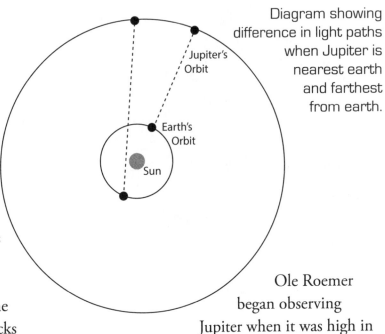

Diagram showing difference in light paths when Jupiter is nearest earth and farthest from earth.

Jupiter's Orbit

Earth's Orbit

Sun

Each hour the clocks differed was 1/24 of the way around the earth. A difference of six hours meant the ship was 1/4 of the way around the world from England. Measuring longitude required that a ship carry an accurate clock, one that kept time throughout the sea voyage. Clocks of the 1600s did not keep time accurately, especially aboard ship with its rocking, swaying motion. They had to be reset often to show the right time.

Astronomers believed celestial events could be used to set the clocks right. But most events in the heavens repeated by the year, month, or day. Those events were more suitable for making calendars. To correct a clock, the event had to take place often and with an accuracy of a minute or so.

The moons of Jupiter seemed to be a solution. The nearest one to Jupiter, Io, traveled around the planet in 42 hours and 28 minutes. On each trip the planet eclipsed it. The navigators aboard a ship would not have to worry about clocks running fast or slow. They could always be corrected by observing the eclipses of Io by Jupiter.

A Danish astronomer, Ole Roemer, worked as Cassini's assistant at the Paris observatory. Cassini had observed the eclipses of the moons and found a troubling problem. They did not come precisely as predicted.

Ole Roemer decided to carefully record the times of its eclipses for a full year. From that he could make a table showing future eclipses. Once they could be predicted with precision, he believed the problem of finding the longitude of a ship would be solved.

Ole Roemer began observing Jupiter when it was high in the night sky. He continued to time the eclipses as earth and Jupiter traveled in their orbits around the sun. After six months, Jupiter was low in the sky at sunrise. It was on the other side of the sun from earth.

But something was wrong. The eclipses were coming later than predicted. He continued his lonely vigil for four years and found a puzzling problem. When earth and Jupiter were at the closest, the eclipses occurred as predicted. When earth and Jupiter were at their greatest separation, the eclipses happened 16 minutes and 40 seconds behind schedule. The error could put a ship off course by more than 250 miles.

On average, however, the eclipses came at the interval predicted. But what good was the average if he could not explain the delay of the eclipses? He redid his figures. His figures were correct. Could his clock be running fast and then slow? Clocks ran either fast or slow, but not first one way and then the other. He built an improved clock and the problem persisted.

Finally, he realized the 16 minute 40 second delay could be explained if light needed that time to cross the distance separating earth and Jupiter. Light had a finite speed and did not flash instantly across space.

Harrison's
Chronometer
H5 from
1772

Roemer realized he could calculate the speed of light, something astronomers had been trying to do for years. Speed is measured by distance divided by time. He needed two values: the distance across earth's orbit from one side to the other, and the total time delay for light to cross that distance.

Cassini had calculated the radius of earth's orbit as 87 million miles. The diameter of earth's orbit would be twice that, or 174 million miles. Roemer calculated the time delay of light in crossing earth's orbit as exactly 1,000 seconds. All he had to do was divide 174,000,000 miles by 1,000 seconds and he had the speed of light in miles per second: 174,000 mi/sec. That is lower than the actual value because the diameter of earth's orbit was not known with enough precision. The actual radius of earth's orbit is 93,000 million miles. Using that figure, the speed of light comes out to be 186,000 mi/sec, very close to the accepted value today.

With the delay understood, Roemer could take it into account and make his table of eclipse times. As it turned out, observing the satellites of Jupiter through a telescope from the deck of a rocking ship proved more difficult than anticipated. Navigators did not like using Jupiter's satellites for navigation. In 1773, William Harrison made a clock that could survive the rigors of a sea voyage and keep accurate time. His invention solved the problem of finding a ship's longitude.

Roemer's efforts show how trying to solve one mystery can lead to a discovery that was not even imagined. Who would have thought that timing the eclipses of a moon of Jupiter would lead to revealing the speed of light?

The satellites of Jupiter make it possible to "weigh" Jupiter. The word "weigh" is in quotations to show that we are using everyday speech. A scientist would instead say that a satellite makes it possible to measure the mass of Jupiter. An object's mass is a measure of the amount of material found in it. Suppose two satellites orbit at the same distance from two different planets. If one satellite orbits more quickly than the other, then its home world has a greater mass.

Density is mass divided by volume. Once the mass of Jupiter is known, dividing that by the volume of the planet gives its density. The calculations show that Jupiter has a density of 1.33 g/cm^3. Jupiter weighs 1.33 times more than an equal volume of water. Overall, Jupiter's density is only one-fourth as much as earth's. Jupiter's lack of density confirms that it is mostly made of light elements such as hydrogen and helium. Despite having more than 1,000 times the volume of earth, Jupiter is only 318 times as "heavy" as the earth. Unlike earth, its core does not contain heavy materials such as rocks and iron.

Once astronomers knew the mass of Jupiter, they could also calculate the surface gravity. The gravitational force exerted by Jupiter on a small object on its surface depends on the square of its distance from the center of Jupiter and the mass of Jupiter. If a planet were twice as massive as Jupiter, but the same size, then gravity at the surface would be twice as much. If Jupiter were made of some lighter material but still kept the same diameter, then the surface gravity would be less than normal.

Distance from the center of the planet is an important factor because it is squared. If Jupiter were somehow to expand and be twice as large, but still weigh the same, gravity at its surface would be cut to one-fourth. A person on the surface would be twice as far from the center of the planet, and gravity weakens by the square of the distance.

Jupiter doesn't actually have a solid surface that we can see. Astronomers use the top of the cloud layer as the surface. To calculate Jupiter's surface gravity, they take into account the total mass of Jupiter and the distance of its surface from the center of the planet. The surface gravity on Jupiter is 2.53 times earth normal. A person who weighs 100 pounds on earth would weigh 253 pounds on Jupiter.

The greater gravity does explain why Jupiter has a thick atmosphere. The clouds of Jupiter are made of hydrogen and helium gas. The molecules of a gas are in constant motion. The lighter the gas, the faster the motion. Hydrogen is the lightest gas, followed by helium. When molecules of hydrogen and helium in earth's atmosphere are heated by sunlight, they actually achieve escape velocity. They escape from the grasp of earth's gravity and are lost to space.

Because of its great mass, Jupiter has a higher gravity than earth. The sunlight that strikes it is 25 times weaker, too. Hydrogen and helium molecules don't gain enough energy to propel them to escape velocity. They remain permanently held in place.

The moons of Jupiter have a weaker gravity. Because of their size, they are more like the terrestrial planets than the gas giant they orbit. If they ever had an atmosphere of hydrogen and helium, those gases have disappeared. Ganymede is made mostly of rock and water ice. The surface has craters covered by an icy crust. Callisto has a thin atmosphere of carbon dioxide. Europa has a smooth surface covered by a thick layer of water ice. Io is closest to Jupiter and experiences strong tidal forces. Its surface is marked by an abundance of volcanoes and tall mountains, some higher than those on earth.

The early 1900s marked a turning away from the study of planets. Large telescopes had revealed much about the planets, but also the law of diminishing returns had set in. Astronomers decided that it would be difficult and time-consuming to continue to merely observe the planets. They turned their powerful telescopes to areas rich with possible new discoveries — stars, nebula, and galaxies beyond the solar system.

It wasn't until the Space Race of the 1960s that new information began pouring in about the planets. The leaders in observing the outer planets were space probes *Voyager 1* and *Voyager 2*. Launched within 16 days of each other in 1977, the 1,600-pound spacecraft are still operating today. At the end of 2012 they had been sending back information for 35 years.

Launching Voyager 2, a scientific satellite, to study the Jupiter and the Saturn planetary systems including their satellites and Saturn's rings.

An image of Europa, the smallest Galilean satellite; the bright areas are probably ice deposits, whereas the darkened areas may be the rocky surface or areas with a more patchy distribution of ice.

Voyager 2 was launched first. It radioed back close-up photos of the moons of Jupiter, Great Red Spot, took measurements of the magnetic field, and then traveled on to Saturn, Uranus, and Neptune.

Voyager 1 also took photos of Jupiter, Saturn, and their moons, and made measurements of the magnetic field and conducted scientific experiments. It gained speed by a method called gravity assist. As the probe approached Jupiter, the gravity of that planet caused it to speed up. When it passed on the other side, it was going faster, so it quickly escaped from the grasp of Jupiter's gravity. The speed it lost leaving Jupiter was less than the speed it gained approaching Jupiter. A second slingshot effect of Saturn gave *Voyager 1* enough speed to eventually pass out of the solar system and into interstellar space.

Voyager 2 had a more roundabout route and did not go as fast. But it too is on its way into deep space, although it trails *Voyager 1* by eight billion miles.

Other space probes have visited Jupiter. Most made fly-bys en route to Saturn and other planets beyond Saturn. But the *Galileo* spacecraft went into orbit around Jupiter in 1995. It orbited the planet for seven years and made numerous close-up fly-bys of the moons.

It showed that the conditions on Europa, the second moon out from Jupiter, are not as harsh as on Jupiter or the other moons. Europa has a very thin oxygen atmosphere, a crust of water ice, and bitterly cold temperature (-260° F.) A space probe built to withstand the cold could soft land on Europa for a permanent outpost to watch Jupiter.

Basic facts about Jupiter are found in chapter 6 in the table on Jovian Planets.

Discover

1. **A satellite was a minor court official**

2. **The Great Red Spot**

3. **Europa has a thin oxygen atmosphere.**

Questions

T F   1. Jupiter is the largest of all the planets in the solar system.

_____   2. Light grows weaker by the _____ of the distance.

A B C D   3. The astronomer who discovered the moons of Jupiter was (A. Giovanni Cassini B. Galileo C. Johannes Kepler D. Ole Roemer).

A B C D   4. Which is NOT the name of one of the four large moons of Jupiter (A. Callisto B. Europa C. Ganymede D. Titan).

A B   5. The four large satellites of Jupiter range in size from almost as large as the earth's moon to about the same size as (A. Earth B. Mercury).

_____   6. Why are photos of Jupiter often printed upside-down?

T F   7. Jupiter spins more slowly than any other planet.

T F   8. The Great Red Spot was first seen by Robert Hooke and Giovanni Cassini.

_____   9. Speed is calculated by dividing distance by _____.

A B   10. A ship's navigator needed an accurate clock to measure (A. latitude — distance north or south B. longitude — distance east or west).

A B   11. The planet with the greater density is (A. Earth B. Jupiter).

_____   12. Gravity decreases by the _____ of the distance.

A B C D   13. Ole Roemer observed a delay in the eclipse of Io by Jupiter because (A. his location in Paris was too far north of the equator B. his timing was off due to an inaccurate clock C. of the small size of his telescope D. of the time for light to travel from Jupiter to earth).

_____   14. The two spacecrafts that passed close to Jupiter and have flown into deep space were named _____ *1* and *2*.

EXPLORE MORE

Research the life of Galileo and write a two-page biography that summarizes his astronomical achievements. The Great Red Spot has been under observations for hundreds of years. How has it changed over the years? What do astronomers think is its cause?

Compare hurricanes on earth with the Great Red Spot on Jupiter. How long can hurricanes last on earth? What is the top speed they can reach? How many years have passed since the Great Red Spot was first detected? What is the top speed of its wind? Hurricanes on earth have an eye. Does the Great Red Spot have an eye?

Both the earth and Jupiter have jet streams. How do they differ in speed, direction, and number?

Ole Roemer found one way to measure the speed of light. Its speed has been measured in other ways, too, by individuals such as James *Bradley*, Hippolyte *Fizeau*, Léon Foucault, and Albert Michelson. Research each method and describe how they measured the speed of light. What is the modern value for the speed of light? Given that Jupiter is 483,600,000 miles from the sun, how long does it take for light to travel from the sun to Jupiter?

Saturn

Saturn is one of the most beautiful sights in the solar system. The slow-moving planet glows a soft yellow in the inky black of space. Beautiful rings surround it.

Ancient stargazers did not know about the rings because they can only be seen in a telescope. But they did notice that Saturn never became as bright as the other planets. Its light was a steady but pale, yellow hue. It moved slowly against the background of stars. Saturn needed almost 30 years to make a complete circle of the sky.

Although Saturn is dimmer than any of the other planets at their brightest, Saturn is easily seen. Its brightness varies but little. Whether at it closest to earth or at its most distant, Saturn is so far away that its brightness is hardly affected.

Early astronomers interpreted its slower motion and dimmer light to mean Saturn was more distant than the other planets. They were correct, at least in speaking of the visual planets. The

Explore

1. **Why did Galileo think Saturn had handles?**

2. **What is a shepherd satellite?**

3. **Do only planets have atmospheres?**

telescopic planets Uranus and Neptune are farther away.

In naming the planet, the Greeks considered its measured passage through the sky and its washed-out color. They decided to name it *Kronos*, the father of Zeus. They pictured Kronos as an old man moving kind of slow. Kronos carried a sickle — a tool with a long, curving blade at the end of a handle that he used as a weapon. In the Roman mythology, the character most similar to Kronos was Saturn, the lord of the harvest. He too carried a sling blade, but his was for cutting down stalks of grain at harvest.

Saturn is dimmer than any other planet, but brighter than any of the stars except for Sirius and Canopus. Both of those stars are in constellations near one another. Sirius is in the constellation of Canis Major, meaning Big Dog, and Canopus is in a constellation of Canis Minor, meaning Small Dog. They are Orion's hunting dogs.

Galileo (again!) was the first person to look at Saturn with a telescope. The planet didn't look round like other planets. Instead, two fuzzy blobs of light glowed on either side of it. It looked like Saturn had two handles, one on either side of the planet. His telescope simply didn't give a clear enough image to reveal the nature of the handles. He tried a different lens to give higher

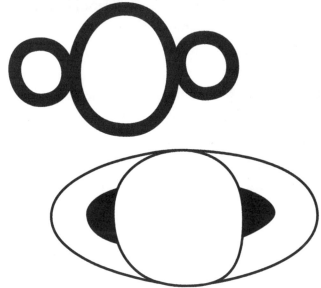

Saturn (above) and its rings have been a beautiful mystery since it was first discovered; at right, two drawings based on Galileo's view of the planet, which appeared to show handles.

magnification. Nothing was gained by making a blurry image twice as big. He still couldn't tell what they were. Galileo studied Saturn a few years later. The fuzzy blobs had vanished. Galileo never figured out what caused Saturn to have handles or why they disappeared.

Fifty years later, Dutch astronomer Christiaan Huygens discovered that telescopes with longer focal lengths gave a better image. His favorite telescope was 27 feet long, although the main lens was only a couple of inches in diameter.

In 1656, Huygens discovered a satellite circling Saturn. Based on its distance and brightness, he believed it was about the same size as the four large satellites of Jupiter, maybe larger. The modern value for its diameter is 3,200 miles, half again as large as earth's moon. Any telescope that shows the four moons of Jupiter will also reveal Saturn's moon Titan. It can be seen in a small telescope at about 50x magnification.

Christiaan Huygens called the moon of Saturn simply Saturn's moon, and it remained unnamed for more than 150 years. But later other satellites were discovered. They needed names to tell them apart. John Herschel, son of William Herschel, who discovered Uranus, suggested the name *Titan* for the largest moon of Saturn.

The Titans were a mythical race of people, brothers and sisters of Kronos (Saturn). In Greek mythology, the Titans were noted for their strength and size. The word Titan is still used today for something that is large such as *Titanic*, the ship, or powerful such as Titans, the Tennessee football team.

Next, Huygens turned to the unusual shape of Saturn. His favorite telescope did not reveal the nature of the "handles." With the help of his brother, he made a larger telescope. The new one was 137 feet long. A tube for it would have been too long and heavy. Instead, he mounted the lens on a tall pole and connected it to his eyepiece with a sturdy cord. This clumsy telescope revealed what Galileo had been unable to see.

He saw that the handles were a ring circling around Saturn. That was such an astonishing conclusion that he wanted to be certain by observing it for a while longer. However, he also wanted to receive credit for the discovery. He wrote a sentence describing his discovery in Latin. Then he rearranged the letters to make a code. He had a book coming out at that time, so he printed the coded sentence in the front of the book.

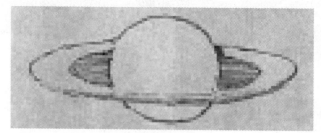

Huygen's discovery of rings around Saturn were so astonishing he delayed announcing their discovery until he could study them in more detail.

Hiding a discovery in code was common practice at that time. Galileo had done it when he first saw the phases of Venus. Scientists would read the code and try to figure out its meaning. Later, the author would reveal its meaning, provided his initial discovery proved correct.

Three years passed before Huygens was certain about the rings. In 1659, he published the decoded Latin sentence. When translated, it read that Saturn was "surrounded by a thin, flat ring, nowhere touching it."

But why did the rings disappear? Huygens realized that the plane of the ring was tipped to that of the earth's orbit. They would be seen edge-on twice during each orbit of Saturn. Although the rings were very broad, they were exceedingly thin. When edge-on they could not be seen in the most powerful telescopes. It was as if they had disappeared. This explains why Galileo lost sight of them.

Christiaan Huygens was a Dutchman who King Louis XIV invited to France. The king had the goal of making France the leading European power. He'd invited Giovanni Casini and Ole Roemer to Paris, too. Louis XIV had invited Huygens to France to be part of an organization of scientists similar to the Royal Society in London. The British scientists came together to communicate new discoveries with one another and pool their knowledge. Only a few months after the establishment of the Royal Society in London, the French Academy of Sciences began meeting in Paris to promote French scientific research in the same way.

Notice the differences in the pictures of Saturn on this page. It seemed to astronomers for a time that Saturn's rings would mysteriously disappear and then re-appear. But the very thin rings are on a plane where they are edge-on and nearly disappear as seen from Earth twice during each orbit of Saturn.

Christiaan Huygens was a Christian who had a strong faith. As the years passed, Louis XIV became more intolerant of Protestants. Christiaan Huygens took his Christian beliefs seriously. By 1681, it had become difficult for him to remain in France and worship God. What should he do? He made the difficult decision to leave the city. He returned to The Hague in the Netherlands. There he could worship God freely.

In 1685, Louis XIV closed all the Protestant churches. He believed that most people would stay in France and worship the way he directed. Instead, more than 200,000 Christians left the country. It was a terrible blow to the economy and scientific advancement of France. Scientists left, as well as many other intelligent, hard-working people.

Huygens continued to make discoveries, but now the honor went to his home country rather than to France.

Today, in amateur telescopes, when a person sees Saturn for the first time, it is an amazing experience. Many other celestial objects can be a disappointment. Photographs of them are made with large telescopes using time exposures and are then enhanced with computers. But Saturn never disappoints. The rings give it a stunning, three-dimensional appearance.

When people first see the planet, they look away from the eyepiece, glance up in the sky at the pale yellow dot, and then back to the eyepiece.

They can hardly believe an object that beautiful actually exists.

The orbits of Saturn and the other planets are almost in the same plane. A plane is a mathematical surface that has length and width but not thickness. Imagine a huge flat surface on which the orbit of earth is drawn. To correctly draw the orbit of Saturn it would have to be slightly above the sheet for part of its orbit and slightly below the sheet for the other part of its orbit. The angle above or below the sheet is called its inclination.

Inclination measures how much each planet's orbit is tilted from the plane containing earth's orbit. Most planets have small inclinations. Their orbits depart from the plane containing earth's orbit by only a small angle. Mercury has the greatest

At left, this photo captures the shadow of Saturn's moon Mimas as it dips onto the planet's rings and straddles the Cassini Division; the close up image of the Cassini Division (right) is an apparent gap in the rings of Saturn

inclination of seven degrees. Jupiter has the smallest, 1.3 degrees, while Saturn is in the middle with an inclination of about 2.5 degrees.

Saturn takes 29.5 years to orbit the sun. For a little over 14 years it is below the plane of earth's orbit and the rings are visible because we are looking down on them from earth. Then Saturn's orbit passes through the plane of earth's orbit. The rings become edge-on as seen from earth and become invisible because they are so thin. But once Saturn gets above the plane of earth's orbit they are visible again. Twice during each orbit, the rings become invisible. The rings stay invisible in an amateur telescope for about 18 months. Saturn presented its rings edge-on in 2009 and at their best in 2015. By 2025 they will be edge-on again.

The tilt of the rings does not shift. It remains constant with reference to the stars. It is merely the shifting position of Saturn above and below the orbit of earth that causes the changing appearance of the rings.

In 1675, Christiaan Huygens learned that Giovanni Cassini had detected a large dark curving line around the ring about a third of the way from the outer edge. Christiaan Huygens thought it was a dark marking on a solid ring. Cassini

believed it to be a gap in the ring. Cassini turned out to be right. The gap he discovered is still called Cassini's Division today.

Slowly the facts about Saturn came in. Isaac Newton in 1687 figured out the density of Saturn. To calculate density, he needed to know the volume of Saturn and its mass. Volume is easily calculated using the formula for finding the volume of a sphere: $V = 4/3(pi)r^3$ with r the radius of the planet — the distance from the surface to the center. Pi is the distance around a circle divided by its diameter. Pi is the same for all circles, about 3.14.

Learning the mass of a planet is especially difficult if it does not have a moon. However, the discovery of Titan solved that problem. Its speed in orbit revealed the strength of Saturn's gravity, and gravity depends on the total mass. The stronger the gravity, the more massive the planet.

Once the mass and volume were known, Newton calculated the density by dividing the mass by volume. In modern terms, the density of Saturn is 0.698 gm/cm³. Saturn is the least dense of all the planets. If a balloon the size of Saturn were filled with water, it would weigh more than Saturn. Stated another way, if an ocean large enough could be found, and Saturn were dropped into it, Saturn would float!

Satellites continued to be discovered. In addition to Titan, Cassini found four more between 1671 and 1684: Tethys, Dione, Rhea, and Iapetus. In 1789, William Herschel discovered two more, Mimas and Enceladus. Mimas is the smallest of the seven with a diameter of 250 miles. Herschel's discoveries gave Saturn more large satellites than Jupiter. Today, both Saturn and Jupiter each have more than 60 known satellites, although some appear to be small objects that were captured as they wandered too close to the planet.

William Herschel also studied the surface of Saturn. Rather than a large red spot as on Jupiter, he found a large white spot. Timing the white spot as it circled Saturn, let William Herschel measure its rotation. Saturn spun on its axis in just over 10 hours. Juptier rotated in just under 10 hours, so

interested in astronomy and built his own private observatory. Later, Harvard University invited him to serve as the university's chief astronomer. They managed to build him a fine observatory with the largest telescope in the world at that time. Its main lens was 15 inches in diameter.

In 1850, while observing the rings of Saturn, he noticed a third, inner ring. It was both dim and thin, because stars could be seen shining through it. It became known as the crepe ring. Crepe is a thin, wispy material. Instead of the ring of Saturn, the correct description is the rings of Saturn.

While all of these discoveries were going on, astronomers continued to debate the nature of the rings. Were the rings solid, liquid, or gas? Some astronomers pointed out that the rings cast a shadow on the planet. Like Christiaan Huygens, they believed the shadow showed that the rings were solid. For 150 years after their discovery,

With giant Saturn hanging in the blackness and sheltering the Cassini spacecraft from the sun's blinding glare, the rings could be photographed as never before, revealing previously unknown faint rings.

they both had very short days.

Because of its low density and high rate of spin, Saturn bulges more at the equator than any other planet. Its equator is 10 percent farther from the center of the planet than the poles.

In 1850, still another ring was discovered by William Bond. He was an American amateur astronomer. He came from a poor family and earned a living as a clockmaker. He became

astronomers tried to find a way to decide on what made the rings.

James Clerk Maxwell solved the mystery of Saturn's rings. He was a mathematician rather than an astronomer. James Maxwell grew up in Scotland in the 1830s. After graduation from Cambridge University in England he became an instructor at a college in Aberdeen, Scotland. His teaching duties did not entirely occupy his time. He looked for a scientific challenge. He learned of a contest known as the Adams Prize. The subject of the contest was "The Structure of Saturn's Rings."

James Maxwell knew that an object in orbit around a planet stays in place because of its speed and the gravity of the planet. Gravity bends the straight-line path into a closed curve. The speed of the object and the force of gravity are in balance. If the satellite goes to fast, it flies off into space. If it moves too slowly, it falls into the planet.

Solid rings, James Maxwell found, would be under tremendous gravitational stress. The rings are 112,000 miles from inner edge to outer edge. Saturn's gravity attracts the inner ring with an intensity four times greater than the outer ring. The inner ring would be urged by gravity to circle the planet at a much higher speed than the outer ring. Even if made of steel, the difference in gravitational attraction would rip the ring apart. The inner ring would fall into the planet and the outer ring would fly off into space. The rings could not be solid.

Could the rings be a liquid? Once again James showed what would occur. Either the rings would break up into individual droplets and escape from Saturn's grasp, or they would stay in orbit but come together to form larger bodies like liquid moons. The rings could not be liquid.

Could the rings be a gas? James knew that a gas is made of molecules in constant motion. Molecules of the same gas at the same temperature do not travel at the same speed. Some travel more quickly, others more slowly. If the rings were made of a gas, the slower-moving molecules would fall into Saturn, while the faster-moving molecules would escape entirely. The rings would be short-lived and soon evaporate.

The rings could not be solid, nor could they be made of a liquid or a gas. What was left?

James Maxwell concluded that many small, solid particles made the rings. Those closer to Saturn traveled the fastest. Those farther out orbited more slowly. Every particle of the ring was an independent satellite of Saturn. They would be small, he believed, perhaps made of small rocks like gravel, or icy objects like hail. Because of Saturn's great distance from earth, the swarms of small particles appear to be solid. James Maxwell won the Adams Prize for his ring study in 1858. (The Adams Prize was named after the mathematician who calculated the location of Neptune. More about him in the next chapter.)

Maxwell's solution caused astronomers to re-examine the calculations made by the French astronomer Edouard Roche in 1850. Roche knew that the earth's gravity caused stress on the moon. Just as the moon causes tides on earth, so the earth pulls on the rocks that make the moon. Gravity grows weaker with distance. The earth pulls more strongly on the near side of the moon than on the far side because the far side is farther away. The difference in gravitational attraction on the near side compared to the far side would be even greater if the moon were near the earth.

Roche wondered if there was a limit to how close the moon could come before earth's gravity would pull it apart. He found that breakup would come should the moon come within 2.44 times the radius of earth; that is, 15,562 miles from the center of the earth. The moon is 240,000 miles away and well outside that limit. It is unlikely to be torn apart by tidal forces.

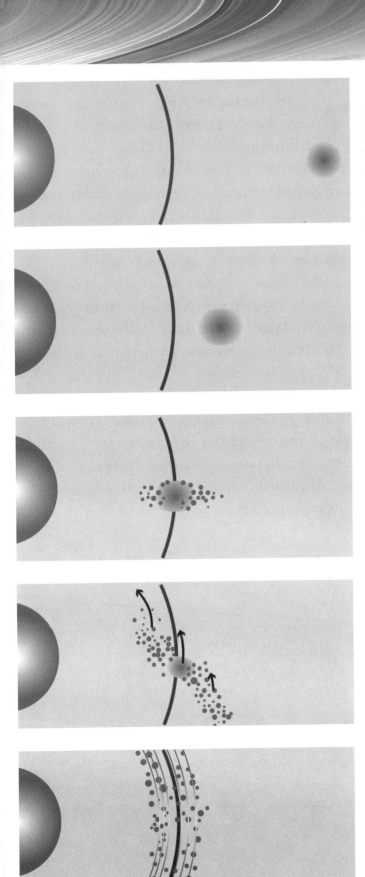

As he worked out the equations, Roche realized that the formula he developed would apply to any planet and its moons. He concluded that satellites would break up if they approached more closely than 2.44 times the radius of their planet. For that reason, 2.44 times the equatorial radius of any planet is called Roche's limit.

His limit appears to be accurate because all moons of all the planets orbit outside his limit, and all known ring systems are found inside the limit. The inner ring of Saturn is only 7,000 miles above Saturn's surface. Some astronomers believe a satellite that gets pulled too close to a planet and then breaks up can cause ring formation.

Tiny objects avoid being broken up because of their size. Each side of the object is about the same distance from the planet. The difference in the gravitational attraction pulling on the side nearest the planet and the side farthest from the planet is slight. The gravel-sized objects that make Saturn's rings are safe from further breakup. Roche's limit does not apply to them.

Once astronomers knew that small particles made the rings, they solved the mystery of Cassini's gap. That particular gap can be explained by looking at the gravitational effect of Mimas, the 250-mile in diameter moon of Saturn discovered by William Herschel. Mimas goes around Saturn in 22 hours.

The distance the moon can get towards Earth before it breaks apart is known as Roche's limit. This equation works for any of the planets and their moons. When the moon approaches the limit point, it begins to be altered by gravitational forces. As it passes the limit, it begins breaking apart and can form rings around the planet from the debris.

Cassini's gap is about 2,500 miles wide and 75,000 miles from the center of Saturn. Should any object stray into the gap, it would revolve around Saturn in 11 hours — one-half of the revolution period of Mimas. Twice during its orbit, a particle in the gap would feel a gravitational tug from Mimas. The gravitation pull of Mimas would eventually draw the object out of the gap. In this way, Cassini's gap is a permanent feature.

The orbits of other satellites can explain other gaps in the rings. Moons that affect the rings and cause gaps in them are called shepherd satellites.

In the 1980s, *Voyager 1* and *Voyager 2* visited the planet. For the first time, questions could be answered that had resisted all efforts to be solved with earth-based telescopes. *Voyage 1* arrived first in 1980 and sent back high-quality photos of the planet, rings, and satellites.

To have a good look at the moons, and especially Titan, mission control changed *Voyager 1's* direction so much that it could not continue on to Uranus and Neptune. Mission control thought it was worth it. During its close fly-by of Titan, *Voyager 1* showed that Titan's atmosphere was mostly nitrogen with a trace of methane.

Titan's atmosphere is thicker than earth's, and its gravity is about the same as earth's moon —— one-sixth earth normal. It would be easy to fly through the atmosphere by strapping on wings and flapping your arms. But you'd need a space suit to protect against the suffocating atmosphere and cold.

In August of 1981, *Voyager 2* continued the study of the Saturn system. The spacecraft showed that the rings do indeed consist of many small particles. They are made of lumps of ice mixed with dust. They range in size from objects the size of small gravel to objects the size of a large automobile. They reflect light well because they are covered with frost. *Voyager 2* used gravity assist from Saturn to gain speed and aim itself toward Uranus.

In 2004, the *Cassini-Huygens* spacecraft took up residence around the planet. *Cassini* had radar to look through the clouds of Titan. It saw several large lakes and their coastlines with numerous islands and mountains. The *Huygens* part of the spacecraft was a lander. It descended onto the surface of Titan on Christmas Day, 2004. After landing, it sent back photos of the surface.

The surface of Titan looks a lot like the surface of Mars. However, the rocks seen in the photos are actually broken slabs of water ice. Because of Titan's -250°F temperature, water ice never melts. The only liquid is methane.

Methane is a type of hydrocarbon, a chemical that is a gas on earth. It is similar to propane, the gas used for heating, but lighter. For it to be a liquid requires a cold enough temperature. Titan is cold enough to turn methane into a liquid. The lakes of Titan are not made of water but of

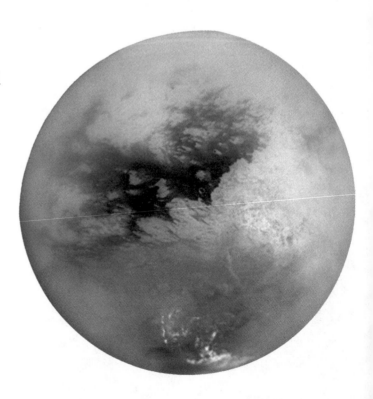

A closer look at Titan, one of the moons of Saturn. The brighter region on the right side and equatorial region of the surface is named Xanadu Regio. Scientists are trying to determine which processes may have created the bizarre surface brightness patterns seen here.

A huge storm churning through the atmosphere in Saturn's northern hemisphere overtakes itself as it encircles the planet as seen from NASA's Cassini spacecraft.

methane. Titan has a methane cycle, as earth has a water cycle. In warmer weather the methane evaporates as a gas, forms clouds, and then falls as methane rain as the temperature cools. A person standing on Titan during that cloudburst would experience a -250°F shower!

Wind and methane rain cause surface features on Titan that look like those on earth such as dunes, rivers, lakes, and seas.

For years, astronomers believed that Titan was the largest satellite in the solar system, even larger than Ganymede. However, over time, a better value for Titan's diameter became available. Ganymede edged it out of first place by 78 miles.

Less is known about Saturn than about Titan. Saturn is about nine times the diameter of earth. The outer atmosphere of Saturn contains about 97 percent hydrogen with the rest helium. Hydrogen is the lightest gas. It used to be pumped into airships, but it proved to be a fire risk. In earth's oxygen atmosphere, a spark can cause it to explode.

Helium is next lightest gas, and for safety it is used to fill balloons.

Saturn's atmosphere has a banded pattern similar to Jupiter's, but more subdued. The colors are not as bright, and it has no really vivid pattern such as the Great Red Spot. Occasionally great swirling storms are sighted, but they are not as long lasting as the storms on Jupiter. One is very regular in that it appears every 30 years at the start of summer in Saturn's northern hemisphere.

The interior of Saturn remains a mystery. Most astronomers believe that in the lower, colder levels where the pressure is greater, the heavier helium becomes a liquid and separates from hydrogen and sinks to the center of the planet. Other astronomers believe the very center of Saturn may have a small rocky core.

Basic facts about Saturn are in the table about Jovian planets in the next chapter.

Another image from NASA's Cassini mission showing the Sotra Facula area of Saturn's moon Titan.

Some of Saturn's Moons

A high-resolution image of Enceladus, another of Saturn's moons, showing the terrain.

A closer look at Herschel Crater on Saturn's moon Mimas reveals there is more than just pure ice, but other dark impurities.

This amazing image of Dione also shows two of Saturn's other moons, Epimetheus and Prometheus, near the planet's rings. Over 60 moons of Saturn have currently been discovered.

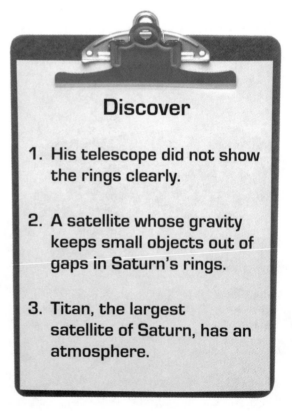

Discover

1. His telescope did not show the rings clearly.

2. A satellite whose gravity keeps small objects out of gaps in Saturn's rings.

3. Titan, the largest satellite of Saturn, has an atmosphere.

Questions

T F   1. The rings of Saturn can be seen in a telescope but not with binoculars or unaided eyes.

A B C D   2. The planet farthest from the sun that can be seen with the eyes alone is _____.
(A. Jupiter B. Neptune C. Saturn D. Uranus)

_____ 3. Saturn is brighter than any of the stars except for _____ and Canopus.

_____ 4. Christiaan Huygens discovered a large moon of Saturn later named _____.

A B C D   5. The person who wrote that Saturn was "surrounded by a thin, flat ring, nowhere touching it" was (A. Galileo B. John Herschel C. Christiaan Huygens D. James Clerk Maxwell).

A B   6. The angle a planet's orbit makes with the plane containing earth's orbit is know as the planet's (A. eccentricity B. inclination).

_____ 7. What is Cassini's Division?

A B   8. Saturn is the _____ dense planet. (A. least B. most)

A B C D   9. The rings of Saturn are made of (A. clouds of gas B. liquid methane
C. small, solid particles D. thin, solid sheets of ice).

T F   10. Roche's limit describes how close a moon can come to a planet without causing tides.

A B C   11. Satellites that cause gaps in rings are called _____ satellites.
(A. shepherd B. Roche C. rogue)

_____ 12. Titan has a _____ cycle, as earth has a water cycle.

_____ 13. The outer atmosphere of Saturn contains about 97 percent hydrogen and the rest is _____.

T F   14. Astronomers know more about the interior of Saturn than they do about its moon Titan.

EXPLORE MORE

Research the life of James Clerk Maxwell and write a two-page biography that summarizes his astronomical achievements.

Water has a density of 1.0 g/cm^3, ice has a density of 0.92 g/cm^3, and Saturn a density of about 0.69 g/cm^3. An iceberg is described as only one-tenth above the surface of the ocean and nine-tenths below the surface. In an ocean large enough, would Saturn sink lower or float higher than an iceberg?

Use a wide-brim hat to illustrate how the rings of Saturn can seem to disappear depending on where Saturn is in its orbit.

Methane is an abundant gas on Titan. How abundant is it on earth? It is a hydrocarbon and has a simple structure made of hydrogen and carbon atoms. Make a drawing showing how the carbon and hydrogen atoms bond to one another. What are some of the other hydrocarbons? What is a hydrocarbon chain?

Write a poem about Saturn.

Jovian Planets

The four planets beyond Mars are known by various names — Gas Giants, Outer Four, and Jovian planets. The word "Jovian" is from Jove, another name for Jupiter. Saturn, Uranus, and Neptune do share similar properties with Jupiter. Each one has a thick and deep layer of clouds made of hydrogen and helium. They rotate quickly considering their great size. They have diameters many times larger than earth's diameter.

Until 1781, no one suspected that any planet orbited the sun beyond Saturn. Ancient astronomers believed that Saturn was the last planet. Because the sun and moon moved across the sky, they considered them planets. That gave seven planets: sun, moon, Mercury, Venus, Mars, Jupiter, and Saturn. From the beginning, seven has been a special number. The week had seven days, and seven is a number that shows completeness. With seven known "planets," no one thought there might be another planet. It would be like wondering if an extra day had hidden itself between Saturday and Sunday.

Explore

1. Who was the musician who discovered a planet?

2. Why is Neptune called a pen and paper planet?

3. What planet is named because of its sea-green color?

Progress

Even after astronomers realized that the earth was a planet, and that the sun and moon were not planets, they never considered the idea of additional planets. The telescope revealed many new objects in the sky. But astronomers didn't look for any more planets because they simply never thought to do so.

Even the person who discovered Uranus was not looking for it. He was a professional musician and an amateur astronomer.

Amateurs have a long history of making important discoveries in science. In astronomy, one of the greatest amateurs was William Herschel. He did not look through a telescope until he was 40 years old. His greatest discovery came while he served as a music teacher and orchestra leader in the resort town of Bath, England, on the Avon River.

William Herschel was born in Hannover, Germany, but moved to England when he was 19 years old. He earned a living by teaching music. As his reputation grew, he played the organ and led the choir of the famous Octagon Chapel. His sister Caroline lived with him, and she sang in the choir. William became conductor of the orchestra, but he kept his many other duties, too.

At age 40, he read a book about astronomy that described the wonders of the solar system. He read about the dazzling white clouds of Venus, the red surface of Mars, the four moons in orbit around Jupiter, and the mysterious rings of Saturn

William Herschel longed to see the wonderful sights described in the books. He checked the price of a telescope. One powerful enough to reveal planetary detail was far more than he could afford.

He rented a telescope for three months. He viewed the skies with the telescope every clear night. Neighbors saw a remarkable sight on nights when he conducted the orchestra. During intermission of the concert he would race out the back way and jump the hedge. He would run down the cobblestone street to his house. Still dressed in his conductor's clothes, he would peer into the telescope set up in his garden. After a few minutes, he would hurry back to finish the concert.

Three months passed quickly, and the day came for him to return the telescope. Rather than satisfying his curiosity, the three months of stargazing only made him want to see more. His goal was to have a telescope of his own. He wanted a bigger and better one than the instrument he had rented. He could not afford to buy one, but maybe he could make one.

The Herschel telescope was built because William Herschel could not afford to buy a large and good quality one – and he hoped to make discoveries with it.

A reproduction of Isaac Newton's rough sketch of a reflecting telescope and its main components

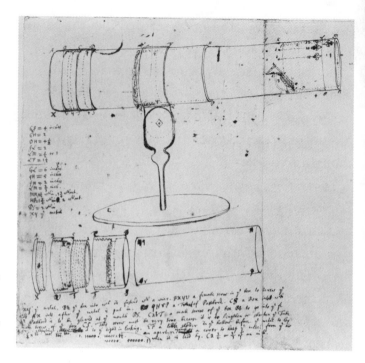

He decided to make a reflecting telescope, which had been invented by Isaac Newton. A curved metal mirror collected light and brought it to a focus to form an image. Reflecting telescopes had mirrors made of a brightly polished alloy known as bell metal. Unfortunately, bell metal tarnished easily and every few months a reflector had to be re-polished. In addition to the extra labor, polishing the mirror had to be done carefully. Otherwise, the polishing changed the shape of the mirror and spoiled its ability to form a sharp image.

William Herschel did not mind the extra effort if it meant he could have a telescope of his own. The first telescope he built was better than none at all, but not as good as he had hoped. However, it was better than the one he had rented, so he knew he could make a better one. As soon as he finished the first one, he began on the second one, and then on a third one. The telescope that met his high standards had a mirror six inches in diameter. The telescope tube was about six feet long, but it was called a six-inch telescope because the mirror was six inches in diameter. With it, Herschel used an eyepiece that gave a magnification of 227.

With the six-inch telescope, he wanted to do more than merely observe. He wanted to make new discoveries. Where should he start?

He knew that professional astronomers seldom spent much time viewing the stars. Stars were very far away. Through any telescope, no matter how powerful, stars appeared as merely points of light. Except as a background grid to show position, professional astronomers ignored the stars. Instead, they studied the planets, moons, and comets in the solar system.

William Herschel decided to make a complete survey of the heavens. He would examine every star visible in his telescope. He pointed his telescope to the sky and let the rotation of the earth bring new stars into view. As the stars drifted across his eyepiece, he examined each one. Even the stars proved interesting. Some were very bright. Others were extremely dim. They varied in color from red to vivid violet. They came in pairs, triplets, and in sprays of stars clustered together.

As he described the stars, his sister Caroline kept notes for him and crossed them off the star chart.

On the night of March 13, 1781, his sky survey took him into the constellation of Taurus the Bull. He found a faint spot of light. It was a dim object of sixth magnitude. He described the star to Caroline.

"It is not on the chart," Caroline told him.

Puzzled, Herschel looked at the object again. Although dim, it was bright enough to be on the chart. He used a more powerful eyepiece. Instead of a star-like point of light, it showed a fuzzy disk. However, going to a higher magnification on a night when the air was not clear and still could cause the fuzzy appearance.

He continued to watch the object. The next night it had moved but only a very small distance.

Hubble Space Telescope view of Uranus showing its four major rings and 10 of its 17 known satellites

The motion was less than that of Saturn, which meant it was beyond Saturn.

Could it be a comet? Comets only became visible when they passed close to the sun. The heat of the sun caused gases to boil off the comet, putting a halo around the head of the comet and causing the long tail to stream away from the sun. No comet had been visible beyond the orbit of Saturn.

Once he made enough observations to calculate an orbit he had another surprise. Most comets travel in long, elongated orbits. They spend much of their time in deep space and then dive toward the sun, cutting across the orbits of Jupiter and earth, and sometimes even pass inside the orbit of Mercury. This object's orbit was nearly circular. Planets traveled in that kind of orbit. And it was in the band of constellations where other planets traveled.

On April 6, 1781, when the night air was especially still and clear, he saw that the object did not have the fuzzy appearance of a comet. Instead, it had a sharp edge like a planet. Herschel announced his discovery, but did not claim it was a planet. He would let the professional astronomers decide. After all, he was merely an amateur.

The object moved like a distant planet. Its orbit was like a planet's orbit. It had a sharp disk like a planet. It was in a constellation through which other planets passed. The professionals quickly agreed that he had discovered a new planet. It was about twice as far from the sun as Saturn. Because of its apparent size at that great distance, it had to be about the same size as Saturn.

What about a name? Herschel called it George's Star after George III, the king of England. Few people outside of England thought that was a good idea. Instead, some tried to name the planet Herschel. Others objected to that name.

Johann Bode, a German astronomer, suggested the name Uranus. Once again, Greek mythology supplied the name. Going outward, Zeus (Jupiter) was the son of Kronos (Saturn) and Kronos was the son of Uranus. Bode suggested the name Uranus for the new planet, and it was accepted.

King George, who had suffered the loss of the American colonies during the American Revolutionary War, was pleased that Herschel had discovered an entirely new world. The king named William Herschel his royal astronomer. The position came with a yearly salary. Now William Herschel could devote his time to stargazing and making improved telescopes. He built one with a mirror 48 inches across. It was the largest in the world. The tube and mounting towered over his house.

He gave up his musical duties. He married and had a son, John Herschel.

He came back to Uranus time and again. In 1787 he noted two satellites circling it. The moons made it possible to calculate Uranus' mass and density. It is about 15 times as massive as earth and its density is 1.27 g/cm^3. It is more dense than Saturn. Rather than floating in a vast lake, it would sink to the bottom.

Herschel noticed that the motions of the moons were peculiar. They appeared to be going over the poles rather than around the equator of Uranus. Unless, that is, Uranus itself was lying on its side. Astronomers confirmed that Uranus was indeed lying on its side. Its axis of rotation was tilted almost exactly 90 degrees. Uranus went around the sun more like a barrel rolling along than a planet spinning upright.

Imagine being at the north pole of Uranus and above the clouds. The sun would be nearly directly overhead on the first day of summer. As the weeks and months pass, it would neither rise nor set, but would instead spiral around and around going ever closer to the horizon. Finally, it would disappear below the horizon. The north pole would have 42 years without sunlight. In the meantime, someone at the south pole would see the sun skimming along the horizon and with each day spiraling higher into the sky until it was almost directly over the south pole.

Astronomers have no explanation as to why Uranus, alone of all the known planets, should be lying on its side.

Fifty years passed before any additional moons were discovered. In 1851, William Lassell, another English astronomer, discovered two more moons. Herschel's son John named all four moons. He chose the names Titania and Oberon for his father's moons. Rather than using characters from Greek or Roman mythology, Titania and Oberon were two characters from William Shakespeare's play *A Midsummer Night's Dream*. The other two he named Ariel, also from a Shakespeare play, and Umbriel from a long poem by Alexander Pope.

The fifth major moon, named Miranda, was discovered in 1948. The five main moons range in size from 293 miles to 980 miles. Titania, in keeping with its name, is the largest. An additional 22 smaller moons have been discovered.

Herschel died in 1819 at the age of 84. By coincidence, 84 years is the time for Uranus to orbit the sun once, its year. His son John Herschel became a great astronomer and scientist, too. Caroline Herschel continued exploring the nighttime heavens. She made discoveries of her own including eight comets. She is often honored as the first woman astronomer. She returned to Germany and died there at age 97.

William Herschel is remembered as the amateur astronomer who discovered a new planet with a homemade telescope set up in his English garden.

Uranus can be seen in binoculars and any small telescope. In fact, a sharp-eyed observer can see it with the unaided eyes. To see the disk of the planet takes a larger amateur telescope at about 300 power and a night with clear and still air.

Voyager 2 was the first spacecraft to visit Uranus. In January 1986 it passed 50,000 miles above the top of the planet's clouds. The images showed that the clouds do not have the bands and storms that are seen on Jupiter and Saturn. Uranus has clouds of hydrogen, helium, and a small amount of methane. The temperature at the surface is -355° F.

Similar to Neptune, Uranus is composed of more ice than gases, and is sometimes labeled as an ice giant.

A multi-year record of Uranus and its rings from 2001 to 2007 as seen from Earth; these Keck Observatory infrared images have south pole at left in the images.

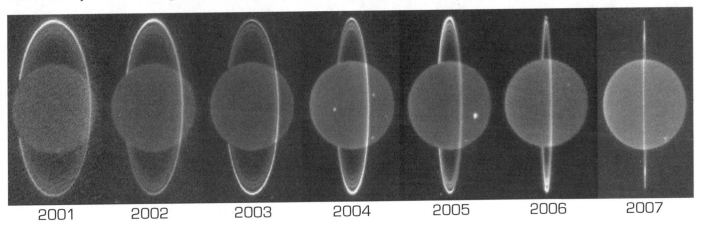

| 2001 | 2002 | 2003 | 2004 | 2005 | 2006 | 2007 |

Because of the low density of hydrogen and helium, something heavier must be below the clouds to give Uranus an overall higher density. The planet may have a core of rocky material, then a layer of icy material of methane and even water ice, and then the known outer layer of hydrogen and helium gases.

Calculating the rotation period is difficult. A portion of the outer clouds appears to be flying around Uranus faster than the planet rotates. The clouds halfway between the equator and poles take 14 hours to go around the planet. But the clouds near the equator take 17 hours.

The other main facts about Uranus are in the table on Jovian planets found at the end of this chapter.

After William Herschel found Uranus, astronomers searched older star charts thinking the planet might have been seen before. Yes, they had noticed it 17 times before. No one recognized it as a planet. Instead, they marked it as a star on their charts and then ignored it.

The older observations did serve a useful purpose. In 1821, French astronomer Alexis Bouvard put together all the observations and calculated an orbit for Uranus. His orbit did not quite match the actual movement of Uranus. He redid the calculations and took into account the slight gravitational attraction of Jupiter and Saturn upon Uranus. Even so, an error remained.

The difference was not much. Uranus strayed about 1/20 of the diameter of the moon from its predicted position. Bouvard concluded that something pulled on Uranus other than Jupiter and Saturn. Mars and the other inner planets were so far away and so small that their gravitational effects could be ignored. Some unseen planet even more distant from the sun pulled Uranus off course.

Bouvard's calculations about Uranus told nothing about the unseen planet. Its size, mass, distance, or what part of the sky the unseen planet might be in was a mystery. Calculating its location would be a real challenge.

In 1841, 22-year-old John Coach Adams tackled the problem. Adams had been born in England and grew up in poverty on a farm. His mathematical skill won him a scholarship to Cambridge University. In addition to his studies, he tutored students and sent money home to his parents. Despite the extra duties, he rose to the top of his class in mathematics, a position known as chief wrangler. Chief wranglers were supposed to tackle difficult problems, so he decided to calculate the position of the mystery planet.

He simplified his task by making some basic assumptions. He figured the unseen planet would travel around the sun in a nearly

These five moons of Uranus are the largest among at least 27 moons for the planet.

Umbriel

Titania

Ariel

Oberon

Miranda

circular orbit. All the other planets had nearly circular orbits, too. He assumed the orbit would take it through the same belt of constellations in which the other planets traveled. He believed this new planet would double the size of the solar system just as Uranus had. Finally, he set its mass equal to that of Uranus.

If John Adams were correct, then in the outer reaches of the solar system a lonely planet orbited. In its sky, the sun would be nothing but an intensely bright star.

Despite his assumptions, the planet could be hiding within a vast starry band that extended all the away around the heavens. His studies and tutoring took most of his time during his school term, but he could work on it full time during vacation. He gathered his reference books and escaped to the solitude at his family's farm in Cornwall. His calculations proved that a more distant planet existed but not its exact location.

He worked on the problem for four years. By September 1845, Adams had the orbit. He knew where to point a telescope to see the planet. His school, Cambridge University, had a telescope with a main lens 12 inches in diameter. It was one of the largest in the world. But the director of the observatory, James Challis, did not think it wise to use precious observing time on what could be a fruitless effort. Instead, he suggested that Adams send the results to George Airy, England's Astronomer Royal. Airy oversaw the work at Greenwich Observatory.

Adams decided to deliver his report in person. However, Airy was in Paris. A month later Adams tried again. The butler answered the door.

"The Astronomer Royal is away but will return later in the day," the butler told Adams.

Adams returned in the afternoon. At last, the Astronomer Royal was in. But — "The Astronomer Royal is eating and cannot be disturbed," the butler said. He refused to invite Adams inside.

John Coach Adams was intensely disappointed. He'd spent four years in his calculations, yet Airy thought his meal more important than a new planet. Discouraged, John Coach Adams returned to Cambridge without ever having met George Airy.

But Adams did leave his report with the butler, and the Astronomer Royal did finally scan through it. George Airy wrote a reply to Adams. Rather than excitement at the possibility of a new planet, Airy expressed doubts about the young student's original assumptions.

The questions baffled John Coach Adams. Then he remembered that years earlier George Airy had written a book about Uranus. In the book, Airy stated his belief that Newton's law of gravity did not hold true so far from the sun.

Adams was astonished that a professional astronomer would question the law of gravity. It had been proven in dozens of different ways. Satellites orbited planets in keeping with the law of gravity. Halley's comet swept far out into deep space, yet it followed the law. William Herschel had found that distant double stars circled one another exactly as Newton's law of gravity predicted.

Unknown to Adams and Airy, French astronomer Joseph Le Verrier had taken up the problem, too. Le Verrier's father had recognized his son's ability at an early age. Le Verrier's father believed in his son's ability so much, he sold the family home to pay for his son's education. Joseph Le Verrier graduated from École Polytechnique, a famous school in France. He stayed on at the school to teach astronomy.

Le Verrier began the calculations necessary to find the missing planet. The mathematics were mind numbing. Six months of effort left him discouraged. Time and again he dropped the problem, only to take it up when he had a fresh idea.

During July 1846, Joseph Le Verrier arrived at an orbit for the unseen planet. He pinpointed its position in the sky. He sent the results to George Airy. The Astronomer Royal suddenly became alarmed. He realized there might be something to Adam's work. The French astronomer and the young English mathematician had arrived at the same conclusion. Only Airy knew this. He kept it secret from both Le Verrier and Adams.

In a panic, George Airy ordered James Challis to use the Cambridge telescope to look for the planet. Airy did not tell Challis the reason for urgency.

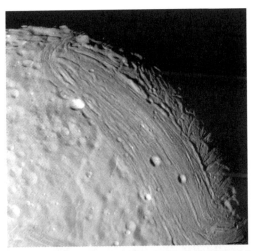

Miranda reveals a complex geologic history in this view, including cratered terrain consisting of rolling hills and craters, a grooved-looking terrain, and one of intersecting ridges and troughs.

James Challis did not act quickly. He thought the planet would be too small to show a disk like a planet and instead look like a star. Finding the new planet would take time. He made excuse after excuse to avoid looking for the planet. When he did begin, he tediously plotted each star he saw. He would see if one moved from night to night. Even at this, he missed the discovery because he actually saw the new planet and marked its position, but failed to double-check his work for movement, and thought it was a star.

Unaware of Airy's urgent plea for Challis to look for the planet, Le Verrier contacted the director of the Berlin Observatory. The director readily agreed to look for the planet and assigned Johann Galle to the job. Le Verrier explained that the planet would be large and would show a visible disk.

Johann Galle did not delay. On the very night he received the search request, he opened the observatory dome and began looking. In less than an hour, his telescope revealed the new planet. The next night it had changed position against the backdrop of stars.

Johann Galle notified Le Verrier so he could announce the success. Although Galle had been the first to see the planet, he understood that the honor of its discovery belonged to Joseph Le Verrier. Galle wrote, "The planet whose position you have pointed out actually exists."

The discovery of the planet on September 23, 1846, was bitter news for England. France and England were rivals. Airy hastily announced that an English citizen had predicted the planet before the Frenchmen. French astronomers quickly reminded Airy that credit for a scientific discovery goes to the person who made the discovery public. Airy had kept Adam's report hidden. French scientists insisted that Le Verrier alone receive credit.

In the uproar that ensued, some French scientists pressed to name the planet Le Verrier.

Neptune as seen from 4.4 million miles away; the Great Dark Spot is seen near the center of the image.

Because he had discovered the planet, Le Verrier had the final say in giving it a name. He decided to avoid making the controversy worse.

What name should he choose? Galle reported that the planet had a sea-green color. Joseph Le Verrier remembered that Neptune ruled the sea in Roman mythology. He selected Neptune as the planet's name.

Galle, Le Verrier, and Adams received many honors. Galle became director of the Berlin Observatory. Joseph Le Verrier became director of the Paris Observatory. John Coach Adams replaced James Challis as director of the observatory at Cambridge.

After more than 40 years, the controversy of Neptune's discovery died down. Le Verrier and Adams had spent a lifetime in the study of the heavens. They were quite content to share in the discovery of Neptune. As their lives drew to a close, Le Verrier came to Cambridge for the dedication of the new Cavendish Laboratory. The two old friends sat side by side during the ceremony. Afterward they strolled across the campus together and talked earnestly about the wonders of science.

The planet became known as the pen-and-paper planet because it was described first on paper before it was seen through the telescope.

As Adams and Le Verrier predicted, Neptune is never bright enough to be visible to the unaided eye. However, it can be spotted in even a small telescope. Although the disk cannot be seen easily, its sea-green color is noticeable.

Only 17 days after Neptune's discovery, a British astronomer found the first moon. It was named Triton. It is about 1,700 miles in diameter, making it the largest by far of Neptune's 13 known moons. The next largest moon is only 260 miles in diameter.

Triton did make it possible for astronomers to calculate the mass and density of Neptune. Its mass is 17 times that of earth, and it has a density of 1.64 g/cm³. As the density shows, like Uranus, Saturn, and Jupiter, Neptune is a gas giant. Or, at that distance from the sun, it is more likely an ice giant. Many of the gases that make it would be frozen at its temperature of -353°F.

Neptune takes 165 years to orbit the sun once. From the time of its discovery in 1846 until the time that a space probe, *Voyager 2*, first flew by it in 1989, the planet had not made a complete orbit of the sun. *Voyager 2* took photos of Neptune and Triton and then used a gravity assist of the planet to start its immense journey into interstellar space. With *Voyager 2's* visit to Neptune, all major planets of the solar system had been visited by a space probe.

The atmosphere of Neptune contains hydrogen and helium, as does all the Jovian planets. Neptune's interior may also have frozen water, ammonia, and methane. At its very center, it may

have a rocky core, but no one knows for sure. More study is needed.

Look at the table of Jovian planets to see how Neptune compares to the other three outer planets.

The table lets you contrast and compare the properties of the planets. You may want to go back to the table of Terrestrial Planets to see how the Jovian planets compare with Mercury, Venus, Earth, and Mars.

Distance: Neptune is a little over 30 times as far from the sun as the earth (30.1 AU) and almost six times as far from the sun as Jupiter (30.1/5.2 = 5.8).

Eccentricity: The orbits of all four planets have an eccentricity near zero (0.048, 0.054, 0.047, 0.011), meaning that their elliptical orbits are nearly circular. Neptune's path around the sun is closest to being circular.

Inclination of orbit: All four stay very close to the plane of earth's orbit. The planet that departs the most is Saturn.

The length of the year, or the time to travel around the sun once, gets longer with distance from the sun. The outer planet, Neptune, not only has a greater distance to go, but it does so at a slower speed.

The rotation rates are measured in hours and all of the planets spin more quickly than earth. Jupiter, the largest planet, also has the fastest rotation rate (9.93 hours).

Diameter and diameter compared to earth: The Jovian planets get smaller as one goes farther from the sun. Jupiter is the largest gas giant; Neptune is the smallest.

Tilt of axis: Jupiter has a very slight tilt to its axis. Saturn and Neptune have a tilt about the same of the earth: 26.7° and 28.3°, compared to earth's tilt of 23.5°. Uranus, on the other hand, is tilted all the way over to its side, 97.8°.

All four of the Jovian planets are far more massive than the earth; however, their surface gravity is not particularly great. Jupiter has a gravity about two and a half times earth normal. The other three would have gravity at their "surface" (top of the clouds) close to earth normal. Gravity falls off with

The approximate sizes of the planets relative to each other. Outward from the sun (at left), the planets are Mercury, Venus, Earth, Mars, Jupiter, Saturn, Uranus, and Neptune, followed by the dwarf planet Pluto. The illustration is for representative purposes, and are not shown at the appropriate distance from the sun.

Jovian Planets

	Jupiter	Saturn	Uranus	Neptune
Average distance from sun (AU)	5.2 AU	9.6 AU	19.2 AU	30.1 AU
Eccentricity of orbit	0.048	0.054	0.047	0.011
Inclination of orbit in degrees	1.31°	2.49°	0.77°	1.77°
Orbital period in years	11.9 years	29.5 years	84 years	165 years
Rotational period in hours	9.93 hours	10.7 hours	17.24 hours	16.11 hours
Diameter in miles	88,846 miles	74,898 miles	31,763 miles	30,775 miles
Diameter compared to earth	11.1/1.00	9.41/1.00	4.00/1.00	3.89/1.00
Tilt of axis in degrees	3.1°	26.7°	97.8°	28.3°
Mass compared to earth	318/1.00	95/1.00	14.5/1.00	17.1/1.00
Gravity compared to earth	2.53/1.00	1.06/1.00	0.905/1.00	1.14/1.00
Density in g/cm³	1.32 g/cm^3	0.687 g/cm^3	1.27 g/cm^3	1.64 g/cm^3
Albedo	0.52	0.47	0.51	0.41
Magnetic field	yes	yes	yes	yes

distance, and because of the Jovian planet's low density and great size, gravity is not as great as it would be if they were compressed into a smaller space.

Density: All four have a low density, about one-fourth of earth normal. This is not surprising because they are made of thick and deep layers of the light gasses hydrogen and helium while earth is made of far denser rocks and has a metal core.

Albedo (reflectivity): About half the light that strikes them is reflected back in space. The farthest planet, Neptune, has the lowest albedo, making it even dimmer in earth's nighttime sky.

All four have a strong magnetic field. Although it is not in the table, all four have a ring system. However, Saturn is the only one to have a prominent ring system.

Discover

1. William Herschel discovered Uranus while still a music teacher.

2. Neptune's position was calculated before it was seen in a telescope.

3. Neptune has a sea-green color.

Questions

_____ 1. The Jovian planets have a thick and deep layer of clouds made of _____ and helium.

T F   2. Until he was 40 years old, William Herschel earned a living as a college math teacher.

A B C D   3. William Herschel first viewed the heavens through a telescope (A. borrowed from a friend B. given to him by his sister C. at a public observatory D. that he rented).

A B   4. William Herschel used his telescope to study (A. satellites and planets B. stars).

A B C D   5. Professional astronomers agreed that Herschel's discovery was a planet because (A. its orbit was like a planet's orbit B. it had a sharp disk like a planet C. it was in a constellation through which other planets passed D. all of the above).

T F   6. William Herschel wanted to name his planet after George III, the king of England.

A B C D   7. The largest moon of Uranus is named (A. Ariel B. Miranda C. Oberon D. Titania).

_____ 8. William Herschel's sister Caroline discovered eight _____.

A B C D   9. John Coach Adam's mathematical skill at college earned him the title (A. arithmetique majeur B. chief wrangler C. geometria master D. prince of mathematicians).

A B C D   10. To simplify calculations, Adams set the mass of the unseen planet beyond Jupiter equal to the mass of (A. the sun B. earth C. Jupiter D. Uranus).

T F   11. Joseph Le Verrie was a French astronomer who also calculated the position of the unseen planet beyond Uranus.

A B C   12. The first astronomer to see Neptune was (A. George Airy, the Astronomer Royal B. James Challis of the Cambridge Observatory C. Johann Galle in Germany).

T F   13. All major planets of the solar system have been visited by a space probe.

_____ 14. The planet that takes the longest to orbit the sun is _____.

A B   15. The density of the Jovian planets is _____ than the density of earth. (A. less B. more)

EXPLORE MORE

Research the life of John Coach Adams or Joseph Le Verrier and write a two-page report that summarizes his main astronomical achievements.

Do all four of the Jovian planets have a ring system? Which planet has the most prominent rings? What do astronomers think are their composition? What causes gaps between rings?

William Herschel made his own telescopes. What metal was used to construct the mirror? What dangers did he face when working with molten metal? What role did his sister Caroline Herschel have in his telescope making? How do his telescopes and mountings differ from small, amateur telescopes today?

Plutoids and Denizens of Space

During the last few years, the planet Pluto has undergone an identity crisis. The International Astronomical Union is an organization that gives space objects their official names. In 2006, they decided that Pluto would no longer be a full-fledged planet. Instead, they put it with smaller objects called dwarf planets. Two years later, they created the name *plutoids* for small, icy planets like Pluto that are found far beyond the orbit of Neptune.

Although Pluto was the first member of the plutoid group, there are others. By 2012, scientists had found ten members of the plutoid family. By then, Pluto had more members in its family than the number of terrestrial planets and the number of Jovian planets combined. New ones are added every year. The number of plutoids may swell to 50 or more.

Explore
1. What outer space object did an 11-year-old school girl name?

2. What discovery resulted from the Bode-Titius sequence of numbers?

3. What outer space objects were believed to be weather events?

4. What outer space objects have a name meaning "long hair"?

Progress

For more than 75 years, astronomers and ordinary citizens thought of Pluto as a major planet like the other eight. Pluto was discovered in 1930, but the search for it began years earlier.

In 1900, astronomers began thinking seriously of looking for a planet beyond Neptune. At first, they looked to see if Neptune stayed on track. But calculating its exact path required observations all around its orbit. By 1900, Neptune was only 56 years into its 165-year orbit since its discovery. They would have to watch Neptune for another 109 years for the information they needed.

What about Uranus? Since its discovery in 1781, it had not only traveled a complete orbit, but by 1900 it was 35 years into the next one. Rather than looking at Neptune, astronomers watched Uranus. Did it follow it predicted path? No, not exactly. Something pulled Uranus slightly out of its proper place.

Percival Lowell, the rich amateur astronomer, had built his personal observatory near Flagstaff, Arizona, to study Mars. His inquisitive mind also turned to the question of whether a lonely planet orbited at the very fringes of the solar system. Percival Lowell was a gifted mathematician who tackled the problem himself. He worked on the problem for five years. Page after page of elaborate calculations could only predict the general part of the sky where the planet might be located.

Lowell's prediction showed that the planet would be far out in space and very dim. His best telescope would show it as a small, dim point of light. He directed his staff to use the telescope to photograph the sky in the target region. The photographic film would collect light from the planet, if it were there, and make it brighter. Lowell examined the photographs personally. For hours each day, he sat at a table, hunched over, and examined the glass negatives with a magnifying glass.

He faced a daunting task. As an object becomes dimmer, the number of dim stars can overwhelm it. The heavens have but 20 stars of the first magnitude, and 6,000 stars that can be seen with the unaided eyes. More than six million stars had the same brightness as the predicted planet. The new planet's position was in a region of the sky especially rich in stars.

The search began in 1905. Percival Lowell worked at the task until he died in 1915. Even after his death, he ensured that the search would continue. He directed that part of his fortune be spent in search of the planet. Yet years passed and it was not found.

In 1929, Melvin Slipher, director of Flagstaff Observatory, bought a new telescope that could take photographs of a large portion of the sky at once. It was called an astrograph. The name comes from *astro* meaning "star" and *graph* meaning "to write." He also purchased a new search instrument called a blink comparator. Two photographs of the same region of the sky were taken three or four nights apart. Then they were put in the blink comparator. It had an eyepiece that could show first one negative and quickly switch to the other. Because stars do not move, they appeared in the same place on both negatives. But a planet does move. When viewed through the blink comparator, the dot that was a planet would jump back and forth.

Clyde Tombaugh worked at the observatory. He was a Kansas farm boy who had only a high school education. He had not been able to attend college. A hailstorm destroyed the family's crops. The family needed the small amount he'd saved for college to survive. But he had managed to construct

a homebuilt telescope, and his interest in astronomy came to the attention of Director Slipher. He offered Tombaugh a low-paying job at the Lowell observatory. Tombaugh eagerly left the farm for Flagstaff.

At the observatory, his job was to clean the machinery, sweep the floors, as well as take photographs through the search telescope. He proved so willing to take on any job assigned him that Slipher put Tombaugh in charge of looking for Lowell's planet. It was tedious and eye-straining work. The negatives were on huge sheets of glass — 14 x 17 inches. Some negatives had almost a half million stars.

For hours he sat with his eye to the blink comparator and searched for the elusive dot that moved. For a year Clyde Tombaugh went through pairs of negatives. But he didn't give up.

On February 18, 1930, he found it. A tiny, dim dot blinked back and forth. It was even dimmer than Percival Lowell had predicted. The amount of movement, about one-fourth of an inch, was the right amount for a planet beyond the orbit of Neptune.

Tombaugh's discovery immediately became a sensation around the world. Newspapers carried headlines not only about the planet but also about the farm boy who discovered it. Tombaugh's home state of Kansas offered him a full college scholarship. He eventually earned a master's degree and taught astronomy at New Mexico State University.

Tombaugh urged the director at Lowell to name the planet quickly; otherwise, newspaper reporters would coin a name that might become permanent. Director Slipher agreed, and decided on an unusual way to find a name. He announced a contest and invited people to submit their choice.

Venetia Burney (right), an 11-year-old girl who lived in Oxford, England, heard about the contest as her grandfather read the morning paper aloud. She'd been studying mythology and suggested the name Pluto. In Roman mythology, Pluto ruled the underworld. It was a cold and dark region. Her grandfather took the time to telegraph her choice to the Lowell Observatory.

Director Slipher and the others at Lowell Observatory voted overwhelming for Pluto as the name for the new planet. They were especially pleased that the first two letters of Pluto, P and L, were the initials of Percival Lowell.

Although no one can be sure, most people believe that Walt Disney, who was 29 years old in 1930, decided to name Mickey Mouse's dog Pluto after the planet. Rather than being the mythical ruler of the underworld most people associate Pluto with a spirited, fun-loving pet.

Clyde Tombaugh's plates, which revealed the discovery of Pluto – the left was taken on January 23, 1930, and the right on January 29, 1930. Arrows point to Pluto.

January 23, 1930

January 29, 1930

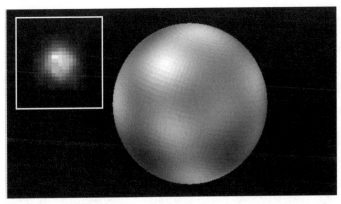

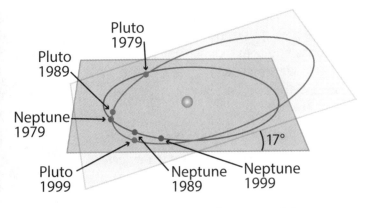

Before seen only as nothing more than a dot of light, the Hubble Telescope image reveals some of the surface of Pluto (left). The dwarf planet has an eccentric orbit (right), that brings it inside the orbit of Neptune. Below Pluto is shown with its moons – Charon, Nix, and Hydra. Two other moons are unnamed.

As astronomers learned more, Pluto seemed an oddball planet. Its orbit was inclined by 17 degrees, far more than Mercury, which has the next highest inclination of 7 degrees. Pluto went around the sun as if on a tilted racetrack, first above the other planets and then below them.

It also had a highly eccentric orbit. Most planets follow a nearly circular path around the sun. Pluto can come as close as 30 AUs to the sun and as far as 49 AUs from the sun. That's a difference of 930 million miles. Between 1979 and 1999, Pluto was closer to the sun than Neptune. During those years it was the eighth planet from the sun rather than then ninth. At that time it was still considered a major planet, although some astronomers were beginning to have doubts.

When first discovered, astronomers believed Pluto would be slightly larger than the earth and have roughly the same mass as the earth. But in 1950, astronomer Gerard Kuiper managed to magnify Pluto enough to see a disk. From the size of the disk he calculated its diameter as 3,600 miles, slightly less than half the diameter of earth and smaller than Mars. In 1965, astronomers predicted Pluto would pass in front of a dim star. The length of time that Pluto hid the star would tell them the size of the planet. When the time came, the star did not wink out at all. Incredibly, Pluto was so small it missed the star entirely.

In 1978, astronomers discovered a moon of Pluto. Using it, they calculated Pluto's mass. Rather than having the mass of earth, it would take 500 Plutos to weight as much as the earth. In 1995, the Hubble Space Telescope showed the diameter of Pluto to be 1,433 miles or about two-thirds the diameter of earth's moon.

Knowing Pluto's mass and size let astronomers figure its density as about 2 g/cm³, less than two-thirds of the density of earth's moon, which is 3.3 g/cm³.

The low mass of Pluto convinced astronomers that it could not have caused changes in the orbit of Uranus. But why did Uranus go off course?

In 2011, there were five different NASA spacecraft gathering data about various planets and areas of our solar system. Four have traveled into deep space beyond the orbit of Pluto.

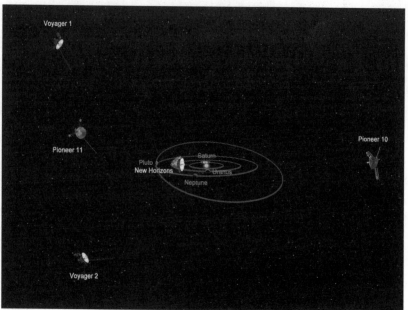

When *Voyager 2* flew by Neptune, the change in the spacecraft's path revealed the reason. Neptune weighed slightly less than first believed. Uranus missed its predicted orbit not because of an unseen planet. Instead, astronomers used the wrong mass for Neptune in figuring how it would affect Uranus.

Pluto had been discovered by chance. It happened to be in the part of the sky where Lowell searched for a planet that turned out not to exist.

Despite its small size, Pluto has five moons: Charon, Hydra, Nix, and two unnamed moons, P4 and P5. Charon has a diameter of 750 miles. It is large compared to Pluto. Rather than Charon orbiting Pluto, they circle around one another, like the weights on the end of a spinning barbell.

Space scientists believe Pluto is made mostly of frozen nitrogen, methane, and traces of other light substances. Learning about Pluto is difficult from earth. Astronomers hope the *New Horizon* space probe will reveal more information. It was launched in 2006 with the goal of reaching Pluto in 2015.

Pluto lies in an outer region of the solar system called the Kuiper Belt. It is named after Gerard Kuiper, the Dutch-American astronomer, who

Eris, a dwarf planet, is show here (center) with its satellite Dysnomia (smaller, positioned 9 o'clock).

studied Pluto and the region of deep space beyond Pluto. The Kuiper Belt is believed to have a large number of comets, dwarf planets, and other small space objects. Those objects similar to Pluto that lie in the Kuiper Belt are called plutoids.

Pluto is but one of many plutoids. Pluto is not the largest of the plutoids. Eris has a diameter of 1,445 miles. The name Eris means discord — a reflection of the arguments that have arisen about what to call Pluto and his kin. Eris orbits the sun at an average distance of 6,324 million miles (more than six *billion* miles), and takes 561 years to complete the trip. Other plutoids include Humea, also discovered in 2003, and Makemake (yes, that's its name), discovered in 2005.

For an object to be called a plutoid, it must be spherical and farther from the sun than Neptune. But plutoids are not the only dwarf planets. Another group, the asteroids, is found between Mars and Jupiter.

The discovery of the first asteroids was the result of a man playing with a series of numbers. In 1766, 15 years before the discovery of Uranus by William Herschel, a German astronomer, Johann Daniel Titius, came up with a list of numbers that expressed the distance of the known planets from the sun. He began with 0, and 3, and then doubled every number after that: 0, 3, 6, 12, 24, 48, and 96. Then he added four to each number: 4, 7, 10, 16, 28, 52, and 100.

He let 10 represent earth's distance from the sun, and calculated the average distance of the other planets using earth as a guide: Mercury 3.9, Venus 7.2; earth 10.0; Mars 15.2. These matched very closely the first four numbers in the list. Then there was a gap for 28 with no planet. But Jupiter 52.0 and Saturn 95.4 were also close matches to his series of numbers.

Another German astronomer, Johann Bode, wrote about the number series in a book he published 1772. Bode felt there was something to the list. Because Bode published it first and called attention to it, the list of numbers became known as the Bode-Titius sequence. Most astronomers saw no particular benefit of the sequence and ignored it.

All that changed in 1781 when William Herschel discovered Uranus. Its distance from the sun, when expressed by setting earth's distance equal to 10, was the next entry in the Bode-Titius sequence. Look at the table "Bode-Titius Law."

Bode-Titius Law			
Starting Sequence	Bode-Titius Series	Relative Distance	Planet
0	4	3.9	Mercury
3	7	7.2	Venus
6	10	10.0	Earth
12	16	15.2	Mars
24	28		
48	52	52.0	Jupiter
96	100	96.0	Saturn
192	196	192.0	Uranus

Titius came up with the sequence of numbers *before* the discovery of Uranus. When earth's distance from the sun is equal to 10, the Bode-Titius series of numbers for the relative distance of Uranus from the sun is 196. Its actual distance is 192.

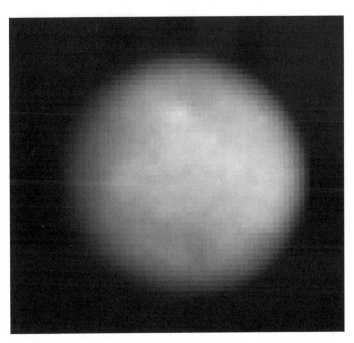
Ceres was originally identified as an asteroid when discovered in 1801, but is now classified as a dwarf planet.

The Bode-Titius list of numbers predicted with an amazing degree of accuracy the actual distance of Uranus from the sun. Astronomers began thinking of the series as the Bode-Titius Law.

It became easy to believe that a planet orbited the sun between Mars and Jupiter to fill the gap at 28. In 1800, 24 German astronomers decided to find the missing planet. They divided the sky into 24 zones, made new and accurate star charts of each zone, and assigned a zone to each astronomer.

While they were planning, an astronomer who was not a member of the search team beat them to the discovery. He was an Italian astronomer named Giuseppe Piazzi.

On January 1, 1801, while working on another matter, Piazzi came across a dim planet. He made a few observations but fell sick before he could calculate its orbit. When he got well, he eagerly returned to the telescope. The little planet was hidden behind the sun. Had he lost it forever? Piazzi's observations were few in number and very inexact. Calculating an orbit was next to impossible.

The giant asteroid Vesta features a mountain more than twice the height of Mount Everest and a set of three craters (circled) known as the "snowman." See the close-up view at right. The largest crater, Marcia, has a diameter of about 40 miles; the center crater, named Calpurnia, is about 30 miles in diameter; and the lower crater, named Minucia, has a diameter of about 14 miles.

Karl Gauss, a 23-year-old mathematician, decided to try his hand at it. Karl Gauss was born of a gardener and a servant girl in Brunswick, Germany. Young Gauss had a fascination with numbers. By the time he was three, he had learned his letters and numbers. His uncle Friedrich saw the boy's genius and convinced Duke Ferdinand to pay for his education at the University of Göttingen.

Gauss tackled the problem of Piazzi's lost planet, as did several astronomers. But Gauss used a mathematical trick of his on invention, called the method of least squares. His solution gave a location where the object would reappear far from anyone else's. On December 7, 1801, astronomers found it at the place where Karl Gauss told them to look.

Piazzi named the object Ceres. Even in a powerful telescope, it was a pinpoint of light like a star. William Herschel suggested that they call it an asteroid, which means star-like. The suggestion described its appearance. The object was not a star but a small planet. Today, it is called a dwarf planet.

Ceres has a nearly circular orbit. It has an average distance exactly at the distance predicted by the Bode-Titus Law, 28; that is, 2.8 AU or about 257 million miles from the sun.

Ceres is 583 miles across. It is small for a planet, but large nonetheless. It has a surface area of about one million square miles. It is as big as Alaska, California, and Texas combined. It is the largest of the asteroids, and the only one whose gravity is strong enough to pull it into a spherical shape. The gravity is about 3 percent of earth normal. A person who weighs 100 pounds on earth would weigh only 3 pounds on Ceres.

A year later, in 1802, a second asteroid, Pallas, was discovered, followed by Juno in 1804, and Vesta in 1807. All three turned out to be smaller than Ceres, although Vesta proved to have a surface that reflected light better. It is the brightest of the asteroids and can be easily spotted in binoculars.

After 1807, no more asteroids were discovered for many years. Astronomers began to think Ceres, Pallas, Juno, and Vesta were all of the asteroids. Then in 1845 came the discovery of another one. After that, new ones came quickly. By 1866, 85 asteroids were known. Today, the number is about 300,000. They are not given a name or number until their orbit has been calculated. One is named Gaussia in honor of Karl Gauss.

The asteroid Gaussia is about 50 miles in diameter. It has a very slight gravity. A 100-pound person would weight about 5 ounces. The escape velocity would be about 20 miles per hour. It would be easy to stand on the surface of Gaussia and put small objects in orbit around it. You could toss a football into orbit. If you hit a baseball with a bat, it could escape from the asteroid entirely.

Several spacecraft have passed near asteroids and taken their pictures as they traveled through the asteroid belt on their way to Jupiter and Saturn. However, some spacecraft have been designed for the sole purpose of exploring asteroids. The American made *NEAR* (Near Earth Asteroid and Rendezvous) orbited the asteroid Eros in 2000, and touched down on its surface in 2001. It was the first spacecraft to land on an asteroid.

The Japanese sent a robot spacecraft named *Hayabusa* (the name means "falcon") to Itokawa, a small asteroid. Itokawa is oblong, about 390 miles by 150 miles. *Hayabusa* was launched in May 2003 and arrived at the asteroid in September 2005. After a month in orbit, it landed and collected a small sample. Then it took off again and returned to earth. In June 2010, the sample pod landed in a desert in Australia. The rest of the spacecraft burned up in the atmosphere.

The spacecraft *Dawn* visited Vesta in 2011. It spent a year orbiting it, taking photographs, and studying the surface. Vesta has a huge crater at its south pole, and the surface is battered and covered with craters, ridges, and grooves. After a year, mission control sent *Dawn* on to Ceres.

Astronomers are fascinated with asteroids because they differ so much from one another. Only a handful of the largest members possess enough gravity to pull them into spherical shapes. Most are misshapen and can be compared more to the shape of a potato than a sphere. They are pock-marked with impact craters.

The most common type of asteroid contains a lot of carbon compounds, which makes them coal black and hard to see. Others are made of metallic iron mixed with silicate-type rocks. Silicate rock contains silicon, oxygen, and a metal such as aluminum or magnesium. Silicates are a common type of earth rock. A few asteroids are rock piles, loose collections of rubble held together by their gravity. Some are composed mostly of the metals iron and nickel. A few are made of water ice and frozen gases mixed with rocks.

Asteroids can be as large as Ceres and as small as — well, astronomers have not yet agreed to the lower limit in size. At first they said that if it was too small to be seen in a telescope, then it was not an asteroid. They set the size at one-half of a mile. As telescopes improved, smaller ones could be detected. Objects as small as 30 feet in diameter can be seen if they come within the orbit of earth's moon. Today, objects smaller than 3 feet in diameter are usually called meteoroids.

The words meteoroid, meteor, and meteorite all refer to the same object. It is a meteoroid when traveling in space, a meteor ("shooting star") when making a fiery plunge through the earth's atmosphere, and a meteorite after falling to earth.

A close view of Eros, an oddly-shaped asteroid with an orbit that takes it close to Earth

A sketch showing the spectacular Leonid meteor shower in 1833

that a Yankee professor would lie than that stones would fall from heaven." He may have been joking, because he knew Professor Stilliman had a good reputation.

Interest in shooting stars increased dramatically on November 12, 1833. As it grew dark, astronomers noted more shooting stars than normal. Then in the early morning hours before dawn, the sky seemed to explode with shooting stars. A thousand meteors a minute shot across the sky. The streaks of light appeared to be coming from the direction of the constellation of Leo the Lion.

People who noticed the display awoke their neighbors, and everyone hurried outside to see the event. At first, many people feared some sort of calamity was taking place. After a while, when nothing bad happened, they began to enjoy the display. The next day, newspapers described the outbreak of shooting stars for those who missed it.

For the first time, astronomers studied meteor showers. They became convinced that meteors came from beyond the earth's atmosphere. A shower followed a path that intersected the orbit of the earth. Meteoroids were scattered all along the path. When earth crossed the meteoroid path, a

Until about 200 years ago, people did not realize that shooting stars were caused by outer space material. They thought they were like lightning or some other weather event. The word *meteor* is a Greek word meaning high in the air. It is the same word found in meteorology, the study of weather.

Some people claimed to have seen meteors actually strike the ground. Most people, including scientists, dismissed the reports. It did seem impossible that rocks could fall from the sky.

In 1807, Benjamin Stilliman, an American chemist, investigated a fallen rock from a meteorite impact. He was a well-respected scientist. Professor Stilliman wrote to President Thomas Jefferson about his discovery. The president was an amateur scientist and a very knowledgeable individual. Jefferson said, "I find it easier to believe

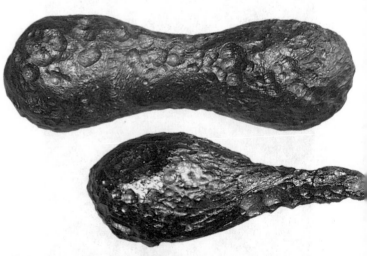

Tektites are usually found in two common shapes - a dumbbell or a teardrop.

meteor shower would happen. Astronomers could predict meteor showers.

Every year in November, shooting stars came from the region of Leo the Lion. The shower became know as Leonids. But meteoroids were not equally spread out along the Leonid's orbit. A cluster of them stayed together. Every 33 years, the Leonids were especially numerous. Rather than a meteor every minute or so, more than a thousand a minute lighted the night sky.

Especially heavy showers became known as meteor storms. The Leonids have not been as vivid lately as they were two hundred years ago. One shower that continues to be strong is the Perseids, which peak on August 12 of each year at over one meteor per minute.

Meteors that are not part of a shower are called sporadic. Sporadic meteors are visible almost every night. They are more likely to hit the earth after midnight because that is when the dark side of earth faces in the same direction the earth is moving. Gravity pulls them in, and the earth's direction of motion sweeps toward them, too.

Although astronomers studied meteors and meteor showers, they did not show much interest in meteorites. Their interest was in things in the heavens, not things on earth. Geology is the study of the earth. But geologists did not study meteorites because the objects fell from space and was not part of the earth.

However, today both collectors and scientists prize meteorites. A successful meteorite hunter can make a living dealing in the precious outer space rocks. They are sold by the gram (454 grams make a pound), and can fetch from 50 cents to three dollars a gram. Unusual ones, or those from a famous meteorite fall, can be worth much more.

Finding them takes skill. Stony meteorites are the most abundant, but also the most difficult to identify because they look similar to ordinary earth rocks. Most stony meteorites contain small, colorful, grain-like spheres. These spheres, called *chondrules*, are a silicate mineral similar to feldspar and olivine. About 90 percent of all meteorites are of the stony type.

Despite their name, stony minerals also contain about 15 percent iron or iron-nickel alloy. Meteorite hunters arm themselves with powerful magnets. If a stony object is attracted to the magnet, then they polish a small section to see if it has chondrules. The presence of chondrules identifies a true meteor.

Iron meteorites are made almost entirely of iron-nickel alloy. They are more easily recognized than other types of meteorites. They are heavy and seem out of place compared to ordinary rocks.

At left, several Perseid meteors as seen in 2012; at right, an image of the Great Comet seen in November 1882

Close Calls

When you think of the universe, you think of a vast expanse with huge distances between stars, planets, and galaxies. However, sometimes an event happens that reminds you just how small the universe can seem.

On February 15, 2013, multiple bright objects were spotted seeming to burn in the atmosphere. Then a small asteroid (estimated to weigh 11,000 tons) streaked across the skies over Russia, traveling at around 50 times the speed of sound. Although believed to be one of the largest objects to enter the atmosphere since the Tunguska event in 1908, which leveled trees in an remote area of Siberia, its unpredicted arrival and its proximity to a populated area, which resulted in some indirect personal injuries and property damage, was shocking and captured the world's attention.

The destruction of the meteorite in the air would leave over 1400 people seeking medical attention, leaving damage to roofs and buildings that had to be evacuated. Overall the event impacted over 100,000 people, caused an estimated damage of over $30 million dollars, and damaged over 7000 structures, including schools and hospital facilities.

The February 2013 event, now known as the Chelyabinsk meteor, is a striking reminder of how much more we need to learn about small, deep space objects that can come within earth's orbit and enter our atmosphere.

Chelyabinsk meteor (above) Tunguska (right)

The third type of meteorites is stony-iron. They are made of almost equal parts iron-nickel alloy and silicate minerals. Some contain larger chondrules made of olivine crystals. They can be cut into thin sections and then polished to a transparent glow. They are the most beautiful meteorites and look like something from another world.

Meteor showers keep a schedule as if following an orbit — a highly elongated orbit similar to those of comets. As more evidence came in, astronomers became convinced that many meteor showers came from the remains of comets.

Let us follow the path of a bright comet. Somewhere out in the Kuiper Belt, or even beyond, a comet floats along in a thousand-year-long orbit around the sun. If left undisturbed, it will be nothing but a collection of frozen gases, ice, and rocks. But something does disturb it — perhaps a passing plutoid, or some other distant object. It is so far from the sun, even a nearby star could upset its orbit. The disturbance sends it heading toward the inner solar system.

It is a small object only about ten miles across. It is light for its size and has a tiny gravity. It you could visit the comet, you would have difficulty walking on it. Each step would send you bounding off into space. If you reached down and gathered up a handful of comet material, it would make a dirty snowball.

In the cold of space beyond Jupiter, a comet remains dark and difficult to see. As the comet continues it journey, an astronomer using a sky patrol telescope may see the faint smudge of light on a photograph. He takes more photos, measures its position, and calculates its orbit. It is large for a comet. Most are about a mile in diameter. But this one is about ten miles in diameter.

Its size and orbit holds promise that it will pass close enough to the earth and the sun to become a bright one. He alerts other astronomers. Reporters learn about it and raise hope of ordinary citizens that they may see a bright comet in a year or so.

The comet passes Jupiter's orbit and falls toward the sun. The pull of gravity from the sun causes cracks to appear in the comet's crust. As the comet passes the orbit of Mars, its surface comes alive. Heat from the sun causes ice below the surface to turn into gas and stream out. A cloud of gas called the coma surrounds the comet. No longer is it ten miles across. Instead, the gas spews out thousands of miles in all directions.

The comet gains a long, ghostly tail. The name *comet* comes from a Greek word meaning "long hair." It becomes a truly bright comet. It has grown to a million miles across and the tail is 60 million miles long. Despite its enormous size, the tail and coma are extremely thin, like wispy fog. Starlight can shine right through them.

The Williamette Meteorite is the six largest known in the world. Though discovered in Oregon, no signs of impact were found.

But the gas reflects sunlight, and it can be seen with the unaided eyes. From earth, the head and coma appear as large as the moon, but with a far softer glow. The tail streams halfway from horizon to directly overhead. Many people come out to look for it. Those who can find a dark sky are rewarded by a spectacular sight. Others must contend with a sky glow from a large city or a suburb filled with streetlights. They find it disappointing. What they see does not compare with the time-lapse photographs of it taken at observatories.

The comet continues its enormous journey. It passes the orbit of Venus and cuts even closer to the sun than Mercury. It whips around the sun, gains speed, and boils away even more of the gas. It is now at its brightest, but no one can see it because of the glare of the sun. A few weeks later it becomes visible once again. It is still a magnificent sight.

As it begins the long journey back, something unusual happens. The tail no longer trails behind

it. The pressure of sunlight and particles streaming from the sun push the tail ahead of the comet. It is as if the comet is facing the sun but backing away from it.

The comet settles down as it grows colder. Gases refreeze, and its tail wastes away. It becomes a faint smudge of light. It may return years later. Or maybe not. The same type of object that disturbed its orbit may act on it again. It may never make the return journey.

Should it return time and again, then in due course it will grow weaker as gases boil away and the rocks that make its core become dislodged. As it disappears, it seeds meteoroids all along its path. They will become shooting stars as a reminder that a comet once made a journey along the same route. When the earth passes close to the remains of the head of the comet, the result will be a spectacular meteor storm.

One of the meteors will make landfall and be found by a meteorite hunter. It will be cut, polished, and worn as a reminder of how beautiful the heavens can be.

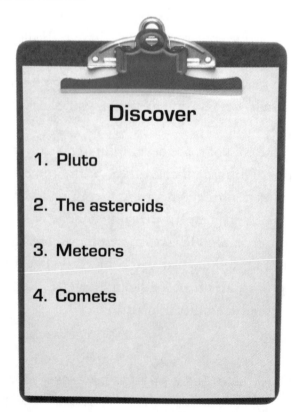

Discover

1. Pluto

2. The asteroids

3. Meteors

4. Comets

Questions

A B C   1. The planetary family with the most members is the
(A. terrestrial planets   B. Jovian planets   C. plutoids).

A B C D   2. The astronomer who first began the search for Pluto was
(A. Johann Bode, B. William Herschel   C. Gerald Kuiper   D. Percival Lowell).

_____ 3. The tool Clyde Tombaugh used to examine photographs in his search for Pluto was called
_____ comparator.

T F   4. Pluto is actually smaller than the earth's moon.

T F   5. Pluto has no known moons.

_____ 6. For an object to be called a plutoid, it must be spherical and farther from the sun than ___.

A B C D   7. The mathematician who computed the orbit of Ceres from a limited number of
observations was   (A. Karl Gauss   B. Giuseppi Piazzi   C. Benjamin Stillman
D. Clyde Tombaugh).

T F   8. The Bode-Titius sequence predicted a planet in the gap between earth and Mars.

A B C D   9. The names Ceres, Pallas, Juno, and Vesta are names of
(A. asteroids   B. comets   C. plutoids   D. satellites of planets).

T F   10. Spacecraft have gone into orbit around asteroids.

T F   11. At first, meteors (shooting stars) were thought to be weather events in the earth's
atmosphere.

_____ 12. Especially heavy meteor showers are known as meteor _____.

A B C   13. Many meteor showers come from the remains of
(A. asteroids   B. comets,   C. an exploded planet).

_____ 14. A handful of comet material would make a dirty _____.

A B   15. As comets repeat their sunward journey time and again,
(A. they grow weaker,   B. they gain in intensity).

EXPLORE MORE

Research the life of Clyde Tombaugh and write a two-page report that summarizes his astronomical achievements.

What is unusual about the eccentricity and inclination of Pluto's orbit? When was the first satellite of Pluto discovered? In addition to Pluto what other objects are found in the Kuiper Belt?

What is the next number after Uranus in the Bode-Titius sequence? Does the distance of Neptune from the sun match the next entry? Most scientists dismiss the sequence as merely a coincidence. What reasons do they give?

How are asteroids different from plutoids? What are some of the methods used for discovering asteroids? What are Trojan asteroids?

Telescopes

The first telescopes used a lens to magnify objects. The first lenses were made for another purpose — as an aid to vision. Many people, as they became older, could not see up close. A convex lens helped a person see near-at-hand objects. A magnifying glass is an example of a convex lens. It is thicker in the middle than at the edges.

In the 1300s, scholars who read books began using optical aid to help them see fine print. At that time books were handwritten, few in number, and very expensive. Not many people could read, and only a few of them needed glasses.

This changed when Johannes Gutenberg invented the printing press in 1450. His printing method used moveable type. Books could be produced quickly and cheaply. As the number of books increased dramatically, so did the number of people who learned to read. Many of these new readers needed eyeglasses.

Lens making became a respected profession. Hans Lippershey ran an optical

Explore

1. How did the invention of the printing press lead to the invention of the telescope?

2. How did Yerkes, Carnegie, Hooker, and Rockefeller help build the largest telescopes in the world?

3. Why are telescopes put on mountains?

Progress

Saint Peter is shown with eyeglasses in this detail of an altarpiece by Friedrich Herlin done in 1466. Eyeglasses had come into use in 1466 when the painting was made but probably were not available in Bible times.

shop in Holland. One day in 1608, an assistant held two lenses to his eye. Some say the assistant was his young son. To his astonishment, a church steeple in the distance appeared closer.

The assistant showed his discovery to Hans Lippershey. The Dutch spectaclemaker instantly grasped the usefulness of the device. He mounted the lenses in a tube and made the first telescope. When officials in the Dutch navy saw the invention, they paid him to make an improved model. The Dutch were a seagoing nation at war with Spain. With Lippeshey's telescope, Dutch ships could spot the enemy before the enemy saw them.

The Dutch did send one telescope to King Henry IV in France. A member of his court had been a former student of Galileo. He wrote a letter to Galileo and mentioned the wonderful invention.

Galileo read the letter and set out to construct a telescope. He gathered lenses and put together a telescope that same day. Galileo's first telescope used a convex lens a little more than one inch in

diameter. Placing his eye to another lens at the other end of the tube showed the image clearly. It magnified three times.

This telescope had only about half the power of an ordinary pair of binoculars.

Rather than using existing lenses, Galileo ground and polished his own. For greater magnification he changed the shape of the lens to project the image farther away. His best telescope had a main lens just under two inches in diameter and formed an image four feet away. It gave a magnification of about 33. This telescope had about the same power as a modern small telescope.

With this improved telescope, he saw mountains, seas, and craters on the moon, the phases of Venus, the moons of Jupiter, as well as additional stars. (He also made the mistake of looking at the sun, and saw dark spots on it. Later, his eyesight failed, and he became totally blind. Never look at the sun!)

Galileo's telescope had two problems. First, it had a limited field of view. It was like looking down a long and narrow tunnel. A large object did not fit in the field of view. Only a part of the object could be seen at any one time. His telescope showed only one-third of the full moon.

The second problem was the lens acted as a prism and produced a rainbow effect. Each color was brought to a slightly different focus. The lens bent blue light more than the other colors and brought it to a focus closer to the lens. The lens brought red light to a focus farther from the lens. Rather than a single image, the lens formed a red image, a green image, and a blue image.

A lens defect was created when Galileo ground down his own lenses; light would enter the lens and act as a prism, which focused the colors differently.

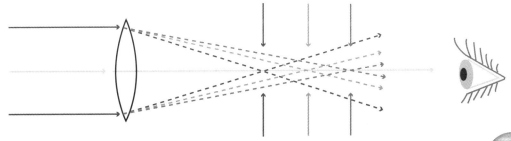

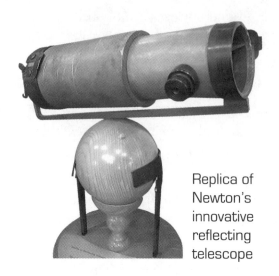

Replica of Newton's innovative reflecting telescope

Each image was nested inside the red image, so a distracting rainbow of color surrounded a bright object.

Johannes Kepler examined telescopes and studied their design. To solve the poor field of view, he switched from a concave eye lens to a convex one. A concave lens is scooped out in the middle. It is thinner in the middle than at the edges. A convex eye lens greatly improved the field of view. It did, however, make the image upside-down. Astronomers preferred the improvement over having a right-side-up view.

Kepler calculated that the confusing color fringes around bright objects could be reduced. His solution used a main lens only slightly curved so it acted less like a prism. It also took more distance to form the image. A lens one inch in diameter needed a tube 17 feet long.

Christiaan Huygens followed Kepler's advice and ended up making incredibly long telescopes. One of them revealed the nature of the rings of Saturn and showed Saturn's moon Titan. Using the long telescopes required skill and patience other astronomers could not muster.

Isaac Newton built an entirely different type of telescope. In 1668, when he was 22 years old, Newton experimented with prisms. He proved that sunlight is made of all the colors of the rainbow. Until his discovery, scientists believed sunlight was pure light and contained no colors. They thought glass caused the colors as sunlight passed through the prism.

Isaac Newton used a mirror to form an image. A mirror reflects all colors the same amount. In his telescope, the main mirror was at the bottom of the tube. Light rays traveled the length of the tube to a concave mirror. The mirror reflected light toward a focus at the top of the tube. Before coming to a focus, Newton put a small, flat, secondary mirror to reflect the rays through an opening in the side of the tube. There it could be magnified further with a regular eyepiece such as the one Kepler designed.

His telescope had a mirror one inch in diameter and a tube about a foot long. Yet it was as powerful as a lens telescope that measured 15 feet long. Astronomers did not immediately accept the idea of a reflecting telescope. It had its own share of problems. The metal used to make the mirror tarnished rapidly. It needed frequent polishing to keep it bright because a tarnished mirror gave a dim image.

Isaac Newton's telescope became known as a reflector, because it used a mirror to reflect light to a focus. A lens telescope is a refractor. Refract means to bend. The lens bends the light to a focus.

In 1758, an English optician named John Dollond discovered how to reduce the disturbing color fringe in refracting telescopes. He put together two lenses, one of crown and the other of

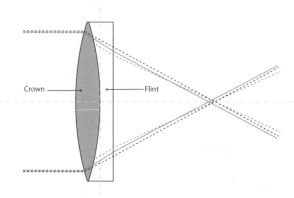

Crown — Flint

A diagram showing the innovation by optician John Dolland to solve the color fringe problem with a refracting telescopes.

William Herschel's telescope towered over the workshop that he used to grind and polish the large 40-inch mirror for his telescope.

flint glass. Flint glass contained lead compounds. It was a dense, scratch-resistant, and very clear glass. Crown glass was common window glass. When light passed through them, they each separated colors by different amounts.

John Dollond made a convex lens of crown glass. It brought rays of light to a focus. He made a concave lens of flint glass. It caused the rays of light to diverge. If the two lenses had been of the same strength, no image would form. But he made the crown lens stronger than the flint lens. The pair of lenses did converge light to a focus, and did so without an overall separation of colors.

Making a telescope with the improved design proved expensive.

Light had to travel through two lenses, so the glass had to be especially pure and free of bubbles. Four surfaces — front and back of each of two lenses — had to be ground and polished. But once finished, the lenses needed no additional attention. Many astronomers preferred the maintenance free refractors over the attention required by reflecting telescopes.

Other astronomers preferred reflectors because they could be made larger. William Herschel discovered the planet Uranus with a reflecting telescope that had a main mirror six inches in diameter. Later he built one 48 inches in diameter. The telescope and mount towered over his house.

Before the modern era, William Parsons, better known by his title of Lord Rosse, made the largest reflecting telescopes. He inherited a large estate in Ireland, and that is where he decided to build the world's largest telescope. He had what it took to make a large telescope — wealth, scientific knowledge, and technical skill. He graduated with honors from Oxford and spent time in London when it was a center of scientific and technical progress.

After building some smaller telescopes, in 1845 he began work on an instrument that become know as the Leviathan. It had a metal mirror six feet in diameter that weighed four

Eyepiece

45°

Plane Mirror

Objective Mirror

How reflecting telescopes work. Light enters the telescopes and travels to the curved main mirror, called the primary. The primary begins the process of bringing light to a focus. Before the focus is achieved, a small flat mirror, called the secondary mirror, is set at a 45-degree angle and directs the converging light outside the tube. A real image is formed that can then be magnified with an eyepiece.

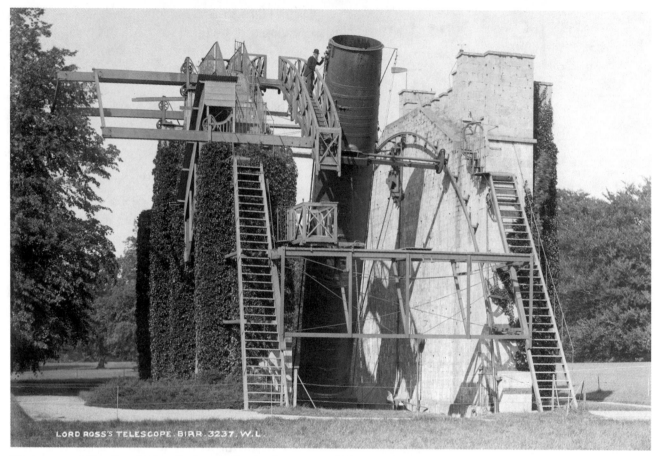

Nicknamed the "Leviathan", the six-foot telescope of William Parsons, Third Earl of Rosse, was located at Birr Castle; at top, is Lawrence Parsons, Fourth Earl of Rosse.

tons. He developed new methods of making large mirrors. He poured the metal in pieces, welded them together, and then coated the blank with molten tin.

Rather than having a gang of men grind the metal to shape, a steam engine did the work. During the polishing stage, the heat generated could cause a mirror to change shape. He did the work under water with the water held at constant temperature. Until then, telescope makers carefully guarded their methods. But Lord Rosse published the details of his successful efforts.

The mirror alone did not make a complete telescope. He also began working on the tube and a way to mount the telescope. He made the tube of wooden planks held together by iron hoops. It was 56 feet long and eight feet in diameter. It looked like a gigantic, long barrel. Mirror and tube weighed 16 tons.

To keep wind from shaking the tube, he built two stone walls 72 feet long and 56 feet high. Once the mirror was in place, the tube could be raised and lowered along a north and south line. However, movement east or west was limited by the stone walls. Lord Rosse let the rotation of the earth bring an object in view. Then he could follow it for about two hours before the tube reached the end of its movement.

His telescope remained the largest in the world until the start of the 1900s. He used it for 20 years, and it was used for another 40 years after his death. In 1908, the tube was lowered to the ground, and it was retired from service. Visitors still come to the site to marvel a size of the Leviathan.

His main achievement was the study of nebulas — what appeared at the time to be great clouds of gas far out in space. His telescope showed that some had spiral shapes. He made a drawing that

Rosse's sketches of the
galaxies M51 (left) and M1

clearly showed the whirlpool structure. Other nebulas, however, had irregular shapes. One nebula in the constellation of Taurus the Bull had glowing filaments running all through it. He came back to it time and again. He gave the object the name Crab Nebula. He never did figure out its nature.

In addition to his astronomical discoveries, Lord Rosse's telescope clearly showed that care should be taken in both the mounting and location of a telescope. The middle of Ireland at near sea level was not the best place to put a large telescope. He lost weeks of observation because the weather did not cooperate. As new telescopes were planned, selecting an appropriate site for the instrument became important. Scouts would travel all over the world to find locations with good seeing — calm, clear air.

A poorly mounted telescope also hindered using it. The telescope needed to turn to the subject

under observation, and smoothly rotate to follow the object to cancel out earth's rotation. Even amateur telescopes of small size need mounts that keep the tube rock steady. At 200x magnification, the slightest wiggle is magnified to be a jar that sends an object careening out of the field of view.

In a telescope, the objective is the lens or mirror that forms the image. The lens that astronomers look through is the eye lens. Telescopes are used to magnify distant objects. Most telescopes have two or three eye lenses, or eyepieces. A telescope user can switch to a different eyepiece to increase magnification. But magnification is limited by the size of the objective. Because of the nature of light, magnification of more than 50x for each inch of the objective is empty magnification. A small telescope with a three-inch lens or mirror has a maximum useful magnification of 150x. Although a more powerful eyepiece can increase the magnification to 300x or more, no additional detail is seen. Greater magnification merely results in an image that is blurry.

Another purpose of a telescope is to gather light to make dim objects bright enough to see. For instance, when Galileo turned his telescope to a group of stars called the Pleiades, he was stunned at what he saw. Sharp-eyed observers claimed to glimpse 7 stars, although most people saw only 6. Rather than 6 or 7 stars, Galileo saw a beehive of stars. He counted 43 of them. It was not the magnification of his telescope that revealed them. Instead, his telescope gathered light and made dim stars bright enough to see.

Light-gathering power of a telescope depends on the area of the objective. Area, in turn, depends on the square of the diameter. For instance, after the human eye gets adjusted to the dark, the pupil, or opening, has a diameter of about six millimeters.

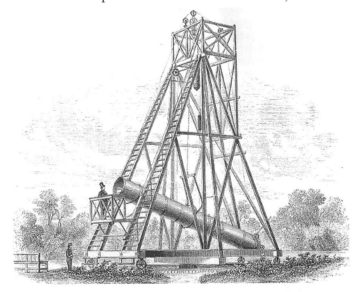

A smaller telescope used by Lord Rosse

Pleiades is an excellent example of why the development and use of telescopes has been so important. Without the aid of such devices, people would only be able to see six or seven of the brightest stars; yet, when viewed with a telescope, dozens more are revealed. In addition, some of the stars are embedded in glowing clouds of gas.

A binocular marked 7x35 has a magnification of 7 and a main lens 35 millimeters in diameter. The binocular will gather about 34 times as much light as the eye alone ($35^2/6^2 = 1225/36 = 34$).

Both magnification and light-gathering power depends on the size of the objective. Telescopes are described not by the length of the tube but by the size of the main lens or mirror. A small telescope for amateur astronomers would have a main lens of about two or three inches in diameter. Mirror telescopes for amateurs range in size from 4.25 inches to 8 inches or more.

Professional telescopes are much larger, of course. Asaph Hall discovered the moons of Mars with a 26-inch telescope. The main lens was 26 inches in diameter.

Much of the progress in building large telescopes came because of the vision of George Ellery Hale. His father made a fortune building elevators for the skyscrapers of Chicago and elsewhere. George Hale was educated at Massachusetts Institute of Technology (MIT), and at the Observatory of Harvard College.

George Hale invented a device to study sunlight and especially solar sunspots. Sunspots come in 11-year cycles. When sunspots are most numerous, the sunspot cycle is at maximum. Slowly the sunspots decrease to a minimum and then return to a maximum in 11 years. Hale proved that sunspots are intense magnetic storms.

Hale became a professor at the University of Chicago. He learned that Alvin Clark had the crown and flint blanks to make a 40-inch lens. Alvin Clark, a master optician, made the best refractors in the world. Both Hale and Clark wanted to turn the glass blanks into the largest telescope in the world. But it would be a costly undertaking.

Hale did his part to raise money. He convinced Charles Yerkes, a wealthy streetcar manufacturer, to put up money for the telescope and a large observatory to be built in Williams Bay, Wisconsin. Clark finished the delicate work of figuring the lenses and fitting them together in 1895. Then it was put in a tube 60 feet long. Despite weighing 20 tons, it could be pointed easily by hand. The floor could be raised and lowered to put the observer at the proper height, regardless of the telescope's position.

When it was first used in 1897, the 40-inch was the largest telescope in the world. It still holds the record as the largest refractor.

George Hale realized that refractors had reached the limit in size. The huge pieces of glass had become heavy, but could only be held by their edges. They sagged under their own weight and changed shape. Light passed through the lens, so the glass had to be perfect through and through. In the case of the Yerkes 40-inch objective, the glass had become so thick it absorbed much of the blue light and reduced its light-gathering ability.

Reflectors, on the other hand, did not have these problems. Light reflected from the surface and did not pass through the mirror. The mirror could be supported from the back so it was less likely to sag. Reflectors could be made more compact and cost less to manufacture.

In the late 1800s, developments made reflectors even more attractive. They could be made of a glass, which was easier to grind, polish, and figure than metal. After being properly shaped, an extremely thin layer of silver was deposited on the glass. When the silver tarnished, it could be chemically dissolved and replaced with a new layer. The process did not change the shape of the mirror, so the mirror did not need to be repolished. Later, telescope makers learned how to use aluminum and other metal alloys to coat the surface of the glass.

The final stage of polishing a mirror is known as figuring. During the grinding and polishing stages, the mirror is automatically given a spherical shape. However, to bring light from a distant object, such as a planet or star, to a sharp focus, the mirror needs to have a parabolic shape. A spherical mirror and a parabolic mirror differ only slightly in shape. Optical rouge — fine, powdery iron oxide — is embedded in pitch and rubbed over the glass. Figuring deepens the center of the spherical mirror into a parabolic surface.

George Hale oversaw building the Yerkes 40-inch refractor, and it came into use in 1897. Only five years later, Hale began making plans for the world's largest reflector. He already had the glass blank that he needed. It came from France. In 1894, Saint-Gobain glassworks cast a 60-inch mirror blank. The original buyer had been unable to pay for it. George Hale's father visited the glassworks. He bought the slab of glass as a gift for his son and shipped it to America. It was made of the same glass used for French wine bottles. The 60-inch disk was 7.5 inches thick and weighed almost a ton.

But a piece of glass alone did not make a telescope. It needed to be ground, polished, and

By 1897, when the Yerkes 40-inch telescope was put in use, refracting telescopes had reached the limit of their size. Any larger and the main lens required a long, heavy tube. The lens itself became so thick its light gathered ability was severely reduced.

figured into shape. A mounting had to be designed that could point the huge mirror to any part of the sky. It would need an observatory dome to protect it. Hale also wanted a location with good seeing. All of this would take money.

George Hale had a unique ability to find funding. In the case of the 60-inch telescope, he asked Andrew Carnegie to pay for both the telescope and an observatory to be built on Mt. Wilson in California. Andrew Carnegie had become fabulously wealthy making steel. He gave away much of his fortune to build libraries, schools, and universities in the United States and other countries. He agreed to pay the cost of building the telescope.

In 1904, George Hale oversaw its construction. The work was done in Pasadena, California. Grinding took two years. Figuring and polishing to the proper shape took two more years. The 22-ton mount was designed to float on a container of the liquid metal mercury so it would move smoothly.

With everything ready, teams of mules hauled the massive parts up the 9.5-mile narrow and winding road to Mt. Wilson. On December 20, 1908, the telescope was ready for use. The 60-inch was pointed to the heavens the same year that Lord Rosse's Leviathan was lowered to the ground.

Although slightly smaller than the Leviathan, the 60-inch proved to be one of the most successful telescopes of all time.

For the first 85 years after its construction, the 60-inch telescope was in use almost every clear night. It was fitted with a spectroscope that breaks starlight into colors and shows the chemical composition of stars. Examination of the spiral nebula that Lord Rosse first sketched showed that the light was similar to that of the sun. Hale believed that it was composed of stars. Rather than a collection of gas, the whirlpool-shaped nebula might be an island universe like the Milky Way itself.

Even today, the 60-inch is still in use. It is the largest telescope in the world fully devoted to public viewing. Amateur astronomers can schedule a half night or full night to use the telescope.

Incredibly, George Ellery Hale had already started a new, larger telescope before the 60-inch was even finished. In 1906, he convinced a wealthy Los Angles businessman, John D. Hooker, to fund the building of a 100-inch telescope.

With Hooker's backing, Hale ordered the Saint-Gobain glass factory to cast a 100-inch blank. The glassworks could not make a complete blank in a single pour of molten glass. Instead, they had to use

The Schmidt camera on Mt. Palomar (above), can take wide-angle photos of the night sky. Sky surveys with it have helped astronomers decide what objects should be given close-up study by larger telescopes such as the 100-inch telescope (right).

two batches. Molten glass pours like taffy, and the two pours produced folds, swirls, and a multitude of small bubbles in the glass.

The 100-inch blank was 12 inches thick and weighed 9,000 pounds. The glass had to be carefully cooled, a process called annealing. It must be cooled slowly over many months. Improper annealing sets up strains in the glass that can cause it to crack or even shatter. The 100-inch blank took a year to cool, and then they sent the blank to Pasadena for inspection.

The 100-inch blank arrived in Pasadena on the same day that the 60-inch mirror was placed on its mounting at Mt. Wilson.

Hale's chief optician was George Ritchey. He inspected the glass and threw up his hands at its condition. It would never do, he said. Ritchey was a cautious individual, nervous, and high-strung. He knew an improperly made glass blank could give all sorts of problems during the grinding, polishing, and figuring stages. The swirls and folds might be evidence of strains in the glass. It might shatter. A piece of glass around the bubbles could chip off and cause deep gashes in the carefully polished surface and ruin months of work.

Ritchey wanted nothing to do with that 100-inch mirror blank. The Saint-Gobain glassworks agreed to try again. The second attempt was even worse. The blank cracked while cooling. Another broke a year later. Maybe, they thought, if it were made thinner it could be done in a single pour of molten glass. It would also cool throughout more readily. The carefully poured blank had none of the defects of the earlier ones — except the blank was so thin it would not hold its shape. Then World War I interfered, and Saint-Gobain glassworks could not make any more attempts.

In desperation, George Hale took another look at the first glass blank. He called in a glass expert. The bubbles were deep enough they should not come to the surface during grinding, the expert

Postcard featuring the 100-inch telescope on Mt. Wilson

said. They might help relieve the strains within the glass as temperature changed.

But Ritchey did not believe it. George Ellery Hale told Ritchey to go ahead anyway. The optician spent five years grinding, polishing, and figuring the glass mirror. He believed his work was doomed to failure. To work so long on what he thought was a futile effort left Ritchey even more nervous and tense.

In addition to the main mirror, Ritchey also made the secondary mirror that sent the light to the side of the tube. In Newton's original telescope, the secondary mirror was no bigger than a fingernail. For the 100-inch telescope, the secondary mirror was two feet by three feet. It took seven months for Ritchey to make it perfectly flat.

Finally, all the work was done. The mirror was installed in its dome on Mt. Wilson. On November 1, 1917, Hale and others gathered to check it out. Ritchey still insisted that the bubbles and internal swirls of glass would render it useless.

They turned the telescope to Jupiter for what they hoped would be an unprecedented look at the planet. What they saw appalled them. Apparently, Ritchey had been right. In the eyepiece, seven dancing images of Jupiter greeted them. The blur filled the entire eyepiece.

Quickly, Hale asked around. Despite instructions, the dome had been left open during the day while workmen put finishing touches on the telescope. Sunlight had shown inside the dome, heating it. The rays of sunlight may have struck

Telescopes continue to get larger, including the 200-inch telescope on Mount Palomar in California

the mirror itself. Hale ushered everyone out and said they would return at 3:00 a.m. By then the mirror should have cooled evenly.

Jupiter had set when they returned, so they turned to a star. In the center of the eyepiece was a brilliant, precisely focused pinpoint of light. The Hooker 100-inch reflector proved a great success. For 30 years, it remained the largest telescope in the world. It revealed the nature of Rosse's whirl-pool nebula. It was made of stars. Rather than a nebula — a collection of swirling gas — it was a galaxy containing millions of stars. For the first time, individual stars were photographed beyond our own Milky Way galaxy.

In 1917, the mirror was equipped with a special instrument that could measure the diameter of stars. Betelgeuse in the constellation of Orion the Hunter was the first star measured. The red star proved to be of gigantic size. If Betelgeuse replaced the sun, the red giant would extend out to the orbit of Mars.

Until the 100-inch saw first light, organizing the properties of stars was a mish-mash of conflicting systems. But with the 100-inch telescope, they could be classified into a sequence based on temperature, size, and brightness. The vast majority were main sequence stars and ranged from hot, blue stars to cooler, red stars. Outside the main sequence were unusually large red giants like Betelgeuse or small, intensely hot stars called white dwarfs.

The 100-inch telescope continued to be used until 1986. Light pollution from the ever-growing cities of Pasadena and Los Angles made

it difficult to observe deep space objects. But the telescope got new life in 1992. Computers, digital cameras, and image processing overcame the smog and light pollution. The telescope still gathers light today.

What did George Ellery Hale think of the telescope? As usual, he was thinking of something bigger. He selected a site for a new observatory on Mount Palomar. Because of its location surrounded by national forest, he believed it would be less likely to suffer light pollution from growing cities. High on the mountain, at 5,600 feet, the atmosphere was clear and stable. It would be a great location for a 200-inch telescope. Such a mirror would be more than 17 feet across — as big as a large living room.

Hale knew every problem he'd had with the 100-inch telescope would be many times more difficult with a 200-inch telescope. The large mirror would weigh much more, slight temperature variations would cause the image to dance around, and it would be so flexible that pointing it to different parts of the sky might cause the glass to change shape. The precision figure Ritchey would produce would be useless.

George Ritchey had a personal solution to the problems. He would have nothing to do with making a 200-inch mirror. He simply could not endure another battle with a large slab of glass. Ritchey and Hale parted company.

Once again, Hale managed to find money to pay for his immense project. This time he turned to the Rockefeller Foundation, set up by John D. Rockefeller, the founder of Standard Oil Company, now called Exxon. The organization concentrated on providing money for hospitals, schools of nursing, developing vaccines, and stopping diseases around the world. Somehow, George Ellery Hale convinced the organization to help build the 200-inch telescope.

Hale learned of Pyrex, a new type of glass developed by Corning Glass Company. Temperature

THE·PRIME·FOCVS·
TWO·HVNDRED·INCH·TELESCOPE

An accomplished man as well as an amateur telescope designer, Russell Porter was brought into a project to build a 200-inch telescope to be placed in an observatory on Mount Palomar. His drawings, like these two examples, were incredibly visionary, accurate, and detailed. They served as a realistic depiction of what the project's blueprints would eventually create. His style of using cutaways and the precision of his mechanical drawings are in a class all their own.

changes caused it less problems than ordinary glass. Hale gave the job of making the blank to Corning, which cast the glass in 1935. The mold had 36 raised blocks in back. The blocks gave the mirror a honeycombed back that reduced the weight from 40 tons to 20 tons.

Hale knew the optical shop at the California Institute of Technology, Caltech, could build the mirror. But he wanted the tube, mounting, and dome to be first rate, too. He asked Russell W. Porter to join the team. Porter was an American engineer, artist, and amateur astronomer. Porter designed the dome and made cutaway drawings of the intricate design of the mounting. His work changed the 200-inch from merely a useful telescope to one of the most beautiful scientific instruments ever built.

A horseshoe-shaped mounting allowed the tube to swing directly to the North Pole and all the way down to the horizon. The primary mirror was so huge that the astronomer could actually sit inside the tube at the prime focus and operate the camera or other instruments in an observer's cage. Other mirrors could direct the light back through a hole in the center of the mirror to other instruments.

The telescope was named the Hale telescope in George Ellery Hale's honor. George Hale did not live to see it finished. He died in 1938. However, he holds the record as the greatest telescope-maker of all time. He built the world's largest telescope four times: the 40-inch Yerkes refractor at Williams Bay, Wisconsin; the 60-inch on Mt. Wilson,

Telescopes on top of Mauna Kea in Hawaii

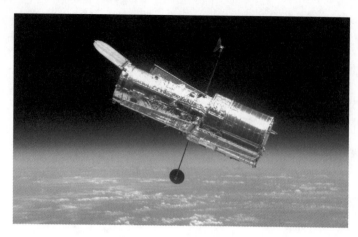

California; the 100-inch also on Mt. Wilson; and the 200-inch on Mount Palomar in California.

The glass giant of Palomar remained the largest telescope in the world from 1949 until 1996. That year, the Keck I telescope on Mauna Kea in Hawaii became the world's largest. It and a twin built a few years later currently hold the record. The Keck telescopes are named for Howard Keck, who put up the money for their construction. Keck I and Keck II are each 33 feet in diameter — almost 400 inches. They are located near the 13,600-foot summit of Mauna Kea in Hawaii.

One of the best-known reflectors is not located anywhere on earth. The Hubble Space Telescope is in orbit 600 miles above the earth. It has a mirror about eight feet in diameter. The space shuttle *Discovery* placed it in orbit in 1990.

Then astronomers made a heartbreaking discovery. The space telescope, which cost nearly two billion dollars, had a serious design flaw. The best image quality was drastically poorer than expected. Experiments revealed the main mirror had been very carefully and precisely figured to the wrong shape. In a serious lapse of quality control, the finished telescope had not been properly tested before being launched into space.

It took three years, but opticians puzzled out a solution. They built a correcting mirror for astronauts to install in the light path of the space telescope. The correcting mirror had the same wrong figure but in the opposite direction.

Astronauts visited the Hubble Space Telescope in December 1993 and installed corrective optics. They continued to service the telescope over the next 10 years and improved it until it was better than first planned. In May 2009 it received its final shuttle mission before the shuttles were retired. In 2010, it celebrated 20 years in space, and it was expected to continue to work for another five years.

Even amateur astronomers were allowed to use the telescope. In all, 13 different sessions were granted to amateurs. They studied why Io, the moon of Jupiter, appeared to change in brightness after being eclipsed by Jupiter. Others studied comets and asteroids.

Professional astronomers used Hubble to photograph southern lights (aurora) on Saturn, carefully measure how variable stars change in brightness, and investigate black holes in distant galaxies. The public also enjoyed some of the truly beautiful full-color images of gas clouds and star clusters made by the telescope.

Discover

1. Reading glasses could be used to make a telescope.

2. They donated money to build telescopes.

3. For clear air away from light pollution.

T F   1. Galileo visited Hans Lippershey to learn how to build a telescope.

_____ 2. When viewed with a telescope, the color fringe around a bright object was caused by the lens acting as a _____.

A B   3. The type of telescope Galileo built is called a (A. reflector B. refractor).

A B   4. To reduce the color fringes, a lens telescope had to be made extremely (A. long B. short).

T F   5. Another way to reduce the color fringe is to make the main lens of crown and flint glass.

A B C D   6. The Leviathan telescope was built by (A. George Hale B. William Herschel C. William Parsons, Lord Ross D. Russell W. Porter).

T F   7. A mirror reflects all colors the same amount.

_____ 8. In addition to magnification, another purpose of a telescope is to _____ light.

T F   9. Twelve years after the 40-inch telescope at Yerkes Observatory was built, a telescope in Paris became the largest refracting telescope in the world.

A B   10. George Ritchey believed the 100-inch glass blank would not make a good telescope mirror because (A. of bubbles and strains in the glass B. it was far too thick).

A B C D   11. The 60-inch and 100-inch telescopes were installed in an observatory (A. in Ireland B. on Mt. Palomar in California C. in Williams Bay, Wisconsin D. on Mt. Wilson in California).

A B   12. For a time beginning in 1986, the 100-inch telescope could not be used because of (A. an improperly figured mirror B. light pollution).

T F   13. The 200-inch Hale telescope is the largest in the world.

A B   14. The Hubble Space Telescope was put into orbit with (A. a serious design flaw B. only enough fuel in the steering jets to last three years).

EXPLORE MORE

Some of the modern observatories include McDonald Observatory in Texas, Kitt Peak National Observatory in Arizona, and the W.M. Keck Observatory in Hawaii. Choose one of these observatories, or another that interests you, and learn about the telescopes in use at the observatory. Describe some of their observing programs.

What is a radio telescope? Read about the early history of radio astronomy. Some of the largest radio telescopes are Arecibo Radio Telescope in Puerto Rico, the Very Large Array in New Mexico, and the world's largest fully steerable radio telescope at Green Bank, West Virginia. Investigate one of these radio telescopes and describe how they are used and what has been discovered with them.

What is a Schmidt Camera? What is a rich field telescope? What is a cassegrain form of reflector? What is the coude focus? How are telescopes used with modern equipment such as Charged Coupled Devices (CCDs) and digital computers to increase their effectiveness?

What is an f/number? The Yerkes 40-inch reflector has an f/number of f/15. The Keck 1 and Keck 2 Telescopes have f/numbers of f/1.75. Why can reflectors be made with smaller f/numbers? What is the advantage and disadvantages of smaller f/numbers?

Chapter 9

Breakthroughs in Astronomy

A breakthrough in astronomy can be a discovery that turns scientific thinking in a new direction. Galileo's discovery of the four large satellites of Jupiter caused astronomers to quickly accept the sun-centered planetary system. Before then, they objected that the earth could not move because it would run away from its moon. But Jupiter orbited the sun and did not run away from its satellites, so the earth could orbit the sun, too, with the moon in tow.

A breakthrough can also be a scientific law that summarizes a vast number of observations. Isaac Newton's three laws of motion and his law of universal gravitation are breakthroughs. Those four laws explained so much about force, gravity, mass, and motion that they became the basis for calculations about moving objects for the next four hundred years.

Explore

1. Who were the "giants" of science?

2. Who became known as the Tycho of the south?

3. Why were astronomers seeking to measure the parallax of a star?

However, scientific discovery is not always measured by an abrupt change, or a great leap forward, or a sudden inspiration. Seldom is an important discovery the work of a single person acting alone. Most scientific advances are steppingstones. The work of one scientist builds on the work of others. Isaac Newton recognized this fact. He said, "If I have seen farther it is because I have stood on the shoulders of giants."

One of the giants Newton referred to was Johannes Kepler, who developed the three laws of planetary motion. Kepler's three laws of planetary motion, in turn, depended on the observations of the great astronomer Tycho Brahe.

Tycho Brahe is best known by his first name alone. Tycho was a prince of Denmark and intended to become a lawyer like his father. But in 1560 he observed an eclipse of the sun. The sight astonished him. He became even more interested in

Tycho Brahe's mural quadrant was an innovation in measuring the location and movement of celestial bodies

astronomy when he learned it could be foretold so accurately. He switched from law to astronomy. He lived his entire life before the invention of the telescope, but he did not let that stop him from being one of the best astronomers of all time.

He quickly learned that predictions of celestial events were not always as accurate as the eclipse he had observed. In 1563, astronomers said Jupiter and Saturn would be in same line of sight as seen from the earth. Their closest approach happened more than a month after the date predicted.

Tycho learned that astronomers still relied on the work of Ptolemy, an ancient Greek astronomer who lived about 150 years after the birth of Jesus. Tycho opened an observatory in Denmark and built instruments that measured star and planet positions better by at least five times. He plotted the position of 777 selected stars. He then used the background of stars to note the location of the planets as their positions shifted from night to night.

He made other observations, too. In 1572, a new star burst into view in the constellation of Cassiopeia. The star grew brighter than Venus and could be seen in the daytime. It continued to burn

A wood cut of the "Comet of 1577"

brightly for a year and a half. Tycho wrote a book about it, *De Nova Stella*, or *The New Star*. The word *nova* for a new star is still used today.

In 1577, a great comet appeared. As Tycho followed it, he realized that its orbit could not be circular. Until then, astronomers believed all motions in the heavens had to be circular, a belief they inherited without testing from the ancient Greeks. In addition, Tycho detected the comet's parallax, which proved the comet had to be farther from the earth than the moon. Until then, most astronomers believed comets were in the atmosphere. The earth's atmosphere grows thinner with altitude. About half of it is below 18,000 feet. Towering thunderheads go up to about 45,000 feet. The highest phenomena seen in the atmosphere are the aurora, northern and southern lights, at a height of 800 miles. Tycho's comet was far above that.

Despite his observations of the new star and the great comet of 1577, Tycho's main goal was to predict where a planet would be at some date in the future. He plotted the position of Mars more often and carefully than any other planet.

Tycho's mathematical skills were not particularly strong. He assigned mathematicians to calculate the orbit of Mars. All of them failed. The accuracy of Tycho's data instantly revealed an error in their work.

In 1597, Tycho moved to Prague, which is today in the Czech Republic but was then part of Germany. Johannes Kepler was a mathematics teacher in nearby Austria. But in 1598 he moved to Prague where he, his wife, and children could worship more freely. There he met Tycho, who was impressed with the young mathematician's superior skill. He invited Kepler to be his chief assistant. Tycho died in 1601 and left his vast observations, what he called his treasure, in the hands of Kepler.

Kepler had been put in change of calculating the orbit of Mars. As had all the astronomers before him, Kepler assumed that Mars traveled at constant speed in a circular orbit with the sun at the center of the orbit.

After six years of painstaking calculations, Kepler had grown desperate. His best orbit misplaced Mars from Tycho's observations by only eight minutes of arc, about the width of a pea held at arm's length. Kepler knew the great Tycho's data did not have that much error.

Finally, Kepler concluded he'd been wrong in all of his assumptions. Mars did not travel at constant speed. It did not travel in a circular orbit, and the sun was not at the center of the orbit. He finally found that Mars traveled in an elliptical orbit with the sun at one at the foci and not at the center of the ellipse. The distance of Mars from the sun did change and its speed did, too. Mars traveled more quickly when near the sun than it did when farther from the sun. Kepler developed three laws of planetary motion.

Kepler's first law states that planetary orbits are elliptical with the sun at one of the foci. The first law described the shape of a planet's orbit.

Kepler's second law states that in the same period of time a line connecting a planet to the sun will sweep out the same area. The second law addresses a planet's speed. Suppose a line connects the planet to the sun. When farthest from the sun, the line is long and the planet travels more slowly.

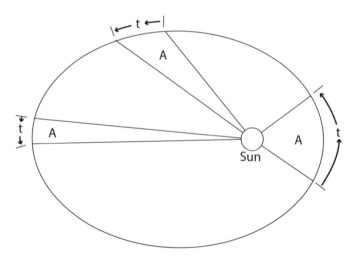

Diagram representing Kepler's second law of planetary motion

When nearer the sun, the line is short and the planet travels more quickly. In the same period of time, the long line moving slowly will trace out the same area as the short line moving more quickly.

Kepler's third law states that the cube of a planet's distance from the sun divided by the square of its period of revolution is a constant and is the same for all planets. The third law related distance of a planet from the sun with its time to orbit the sun. Planets that go around the sun more quickly are closer to the sun, and those that go more slowly are farther from the sun. But the third law was more than a general statement. It was a specific equation: d^3/T^2 = constant, with d the distance from the sun and T the time to go around the sun.

Kepler's third law of planetary motion let astronomers calculate the relative size of the solar system using the earth's distance from the sun as a measuring stick. For instance, the earth takes 365 days, or 1.0 year to orbit the sun. Mars takes 687 days, or 1.88 earth years to orbit the sun: 687/365 = 1.88. If we square 1.88 and then take the cube root, the result is the distance of Mars from the sun in AUs: $(1.88^2)^{\wedge}(1/3) = 1.52$. The symbol ^(1/3) is another way of writing the cube root. Mars is 1.52 times as far from the sun as the earth. Once astronomers found the distance of the earth from the sun, they could multiply that number by 1.52 to find Mars' actual distance from the sun: 93,000,000 x 1.52 = 141,000,000 miles, or 1.41 million miles.

While Kepler was developing his laws of planetary motion, Galileo studied motions of objects on earth. Galileo did experiments with falling objects and with balls rolling down an incline plane and on a tabletop. Galileo found that on a flat surface a ball rolled in a straight line at constant speed unless acted on by friction. Only then did it slow.

Galileo died in 1642, the same year that Isaac Newton was born. Isaac Newton combined the work of Kepler and Galileo and extended it. To most people, it appeared objects in space did not follow the same rules as objects on earth. For instance, the moon kept moving at the same speed around the earth year after year without stopping. A rolling ball soon came to a stop. Galileo said the ball rolled to a stop because of the force of friction.

Isaac Newton summarized Galileo's observation with the first law of motion: If an object begins moving, changes speed or direction, or comes to a stop, then a force of some kind has acted on it.

Galileo also did experiments to measure the speed of falling objects. He found that a falling object gained 32 ft/sec in speed for each second it fell. At the end of the first second its speed was 32 ft/sec, at the end of the next second it was going 64 ft/sec, and at the end of the third second it traveled at 96 ft/sec. The acceleration due to gravity was 32 ft/sec per second, or 32 ft/sec².

Newton's second law says that the acceleration of an object is directly proportional to the force acting on it and inversely proportional to the object's mass. Double the force and an object accelerates twice as fast. But an object with twice the mass needs twice as much force to get it moving at the same acceleration.

Newton's third law of motion states that forces always come in pairs acting in opposite directions on two different objects. As the earth's force of gravity acts on the moon, the moon also acts on the earth. They are equal forces, but the moon does not affect the earth's path as much because the earth is more massive than the moon. But there is a slight wobble to the earth's orbit caused by the moon.

Newton's law of gravity is a force equation, too. On the earth's surface, the acceleration due to gravity is 32 ft/sec². An object, such as a falling apple, is 4,000 miles from the center of the earth. The moon is 240,000 miles from the center of the earth, or about 60 times farther away than the apple. The moon is undergoing acceleration because its straight-line direction is changed into a curve to orbit the earth. Newton found

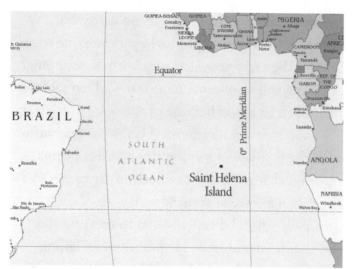

Site of Halley's telescope to explore the universe from the Southern Hemisphere

the acceleration of the moon was NOT 60 times weaker than that of the apple, but 60x60 = 3,600 times weaker. Gravity falls off, or grows weaker, by the square of the distance.

Astronomers tested Newton's law of gravity using Kepler's three laws of planetary motion and found they were in complete agreement. Newton's law of gravity not only explained why planets travel in elliptical orbits, but also why they go faster when near the sun and slower when farther from the sun. In addition, it applied equally well to the satellites in orbit around planets as it did to planets in orbit around the sun.

Isaac Newton enjoyed finding solutions to knotty problems. However, once he found a solution, it no longer interested him. He would put it aside and take up a new problem. He discovered the law of gravity in 1666, but then did not make it public. It lay forgotten for 20 years until Edmund Halley, a young astronomer, visited Isaac Newton with a question about gravity.

Edmund Halley was born in London in 1656. He had studied at Oxford, but he left before finishing school. He had become so interested in astronomy he couldn't wait to become a full-time astronomer. He became the assistant of John Flamsteed, England's first Astronomer Royal.

Flamsteed was busy improving on the work of Tycho Brahe.

But Flamsteed could not observe stars in the southern skies. Southern stars remained below the horizon in England, so Flamsteed could not make an accurate chart of them. In addition, most of the great observatories were in the Northern Hemisphere, so knowledge of the southern skies came from sailors and travelers. There was no good chart of the southern stars.

Edmund Halley decided to do for the Southern Hemisphere what Tycho had done for the Northern Hemisphere. He packed a telescope and traveled to St. Helena, a lonely island of volcanic origin in the middle of the South Atlantic. It was one of the most isolated places on earth.

Halley's primary goal was to precisely locate a network of stars. But he marveled at some of the sights in the southern skies. Centaurus was one of the better-known southern constellations. It was located south of Lupus the Rabbit. Lupus, in turn, was just south of Orion the Hunter. Observers in England could see Orion easily, and could just see Lupus low on the southern horizon. But they could

Omega Centauri is one of the few globular clusters visible to the naked eye and appears about as large as the full moon.

not see Centaurus at all because it was always below the horizon.

At St. Helena, Centaurus was higher in the sky. Halley made a remarkable discovery. One of the brighter stars, Omega Centaurus, was a misty ball of light. It looked like the head of a comet, but it didn't move. Instead, it stayed fixed in place like a star.

More than a hundred years passed before better telescopes revealed its true nature. Omega proved to be a globular cluster. The word globular means shaped like a globe. Omega contained a swarm of 300,000 hot, bright stars. They looked dim only because they were so far away.

Halley also studied two sights seen by the navigator and explorer Ferdinand Magellan. During his sailing trip around the world in 1519–1521, Magellan noted two patches of stars in the southern night sky. The clouds looked as if they had broken away from the Milky Way. They were named the Greater and Lesser Magellanic Clouds in honor of Magellan. Halley studied the Magellanic Clouds, although he did not succeed in understanding their nature. Later, astronomers showed that they are galaxies. They are the nearest neighbors of our Milky Way galaxy.

After two years, Halley returned to England and published his report. He'd carefully plotted the

Seen from vantage point of the southern skies, the Large and Small Magellanic Clouds (LMC and SMC) are bright patches in the sky. They are irregular dwarf galaxies, and along with our Milky Way Galaxy, belong to what is called the Local Group of galaxies. These two features of the night sky are easily seen without the aid of a telescope although the SMC is 200,000 light years away and the LMC is 160,000 light years away.

Within the Large Magellanic Cloud is an abundance of stellar features similar to those found in our Milky Way, including star clusters and large dust clouds.

positions of 341 southern stars. His work earned him the title Tycho of the South. Scientists of the Royal Society recognized Halley's accomplishments. They elected him a member.

After returning from St. Helena, Edmund Halley made his home in London. As the chief city of the British Empire, it was an exciting place to live. Great scientists came together at the Royal Society in London each week to discuss the latest discoveries in science.

One day in 1684, Halley met with Christopher Wren and Robert Hooke. Christopher Wren was an architect and amateur astronomer. Wren supervised the rebuilding of London after the Great Fire of 1666. Robert Hooke was Wren's assistant and a talented experimental scientist. The discussion that day was about gravity.

All three of them knew that as light spread out, it grew dimmer by the square of the distance from the source of the light. They figured that gravity had to act the same, but no one could prove it. Edmund Halley figured that if anyone could answer the question about gravity, Isaac Newton would be the one. Newton lived on the grounds of Cambridge University where he was professor of mathematics. Edmund Halley traveled there and questioned Newton about gravity.

Newton explained that almost 20 years earlier he'd developed an equation that described how gravity grew weaker by the square of the distance. He had even worked backward from the law and developed Kepler's three laws of planetary motion. If gravity fell off by the square of the distance, then the planets would travel in elliptical orbits.

Halley insisted that Newton publish his discovery.

Newton said, "I hesitate to take the time needed to see a book through the printers."

"I'll do all of that for you," Halley promised. "You write the book, and I will take care of the details."

During the next 18 months, Isaac Newton wrote his book, which he called *The Principia Mathematica*. Edmund Halley took care of the details, including paying to have it published out of his own pocket. *The Principia* quickly earned a reputation as the greatest scientific book of all time, and Isaac Newton as the greatest scientist.

While working with Newton on *The Principia*, Halley was struck by the fact that Newton applied the law of gravity to all objects in the solar system, even comets. Of all the bodies in the solar system, comets were the most mysterious. They could appear from any direction. As they fell toward the sun, a great glowing tail formed. They would whip around the sun and then disappear into the far reaches of the solar system, never to be seen again.

Comets were unpredictable and appeared to follow no set rules. Isaac Newton claimed they followed the law of gravity, too.

Halley thought, *If comets obeyed the law of gravity, their paths will follow a predictable course.*

Halley looked for reports of bright comets. He'd seen one in 1682. A bright comet had been seen in 1531 and 1607. Each sighting had been about 76 years apart. He figured the orbits of each one. The orbits were very similar. The difference could be accounted for by assuming the comet had been pulled off course by a large planet such as Jupiter. Halley boldly claimed that it was not three different comets, but the same comet returning repeatedly.

Halley confidently predicted the comet would again pass from the dark reaches of deep space and whip around the sun in 1758. The long-awaited comet came into view on Christmas night in 1758, 16 years after Edmund Halley had died.

Edmund Halley's discoveries did not end with his work with Isaac Newton. While looking through a star map made 1,500 years earlier, he found at least three stars — Sirius, Procyon, and Arcturus — in a slightly different location than shown on modern star charts. He searched through even older star charts. Since the time of the ancient Greeks, Sirius had moved the width of the full moon.

Halley, like other astronomers, believed stars to be fixed in place. Each day they rose, crossed the sky, and set. That motion was due to the rotation of the earth. Their positions compared to their neighbors did not change. But Halley discovered otherwise.

"Stars do move," Halley announced to the scientists of the Royal Society. "Even in the 100 years since Tycho first plotted its location, Sirius has shown a slight proper motion."

Years later, William Herschel added to what Edmund Halley had discovered. Herschel carefully marked lines showing the proper motion of the stars. It appeared that those in the constellation of Hercules were streaming away from one another. He, however, realized that it was due to the sun moving toward

One of our galaxy's closet interstellar neighbors, the SMC is a rich source of study for scientists. It is actually a dwarf galaxy, but it is so bright, it can easily be seen in the night sky.

The Large Magellanic Cloud (LMC) and Small Magellanic Cloud (SMC) are shown at left in this image as the Milky Way dominates the horizon.

them. Not only did the stars move, but so did the sun, and it carried its planets with it.

William Herschel set out to measure the parallax of the nearer stars. From a star's parallax, he could calculate the distance to the star. Parallax is the apparent shift of a nearby object against the background due to a changing viewpoint. The distance between the two viewpoints is the baseline. Earth's orbit around the sun was the largest baseline available to Herschel. Earth's orbit has a radius of 93 million miles and a diameter of 186 million miles.

When Herschel viewed stars six months apart, it was like seeing them with eyes separated by 186 million miles. To improve his chances to detect the slight shift, Herschel looked for pairs of stars that were close together. He assumed they were far from one another — one was closer (the brighter one, most likely) and one was much farther away (the dimmer one). They appeared double merely because by chance they fell in the same line of sight with one behind the other. As the earth traveled around the sun, the closer star should shift back and forth.

By 1784, William Herschel had found 753 double stars. Actually, he found so many

double stars it cast doubt on his assumption that they fell by chance near one another. Twenty years later he came back to his double stars. He found that in three cases, he could detect very slight motion. It wasn't motion due to parallax. They were true binary stars — two stars that orbited one another.

William Herschel never did detect a parallax. He died in 1822.

The astronomer who did succeed in measuring a star's parallax was Friedrich Wilhelm Bessel, a German astronomer. Bessel was born in 1784, about the time Herschel began collecting double stars. Bessel's formal education ended at age 14, and he became an accountant. But he continued to study languages, mathematics, navigation, and astronomy. He came to the attention of professional astronomers when he recalculated the orbit of Halley's comet.

Bessel was invited to become an unpaid astronomer's assistant. His current job as an accountant did not pay much anyway, so he decided he preferred to starve as an astronomer rather than as an accountant. However, by 1809 he proved to be a talented astronomer. Frederick William III of Prussia put him in charge of a new observatory.

He could not accept the post right away because it required that he have an advanced college degree. The University of Göttingen granted him a degree without him ever attending that school.

After nailing down the precise positions of 3,200 stars, he began the project of finding one with a measurable parallax. For this effort he had an unusual telescope. Joseph von Fraunhofer, a master telescope maker, had designed it. Fraunhofer had taken a six-inch lens and cut it in half with a diamond saw. One half he set so it couldn't move, but he let the other half be moved very precisely. The ingenious device could measure angles between stars very accurately.

Bessel first put the two half lenses together so they made a single lens. Then he would focus on two stars whose separation he wanted to measure. Next, he turned a precision screw that slowly slid sideways the movable half lens. This gave him a double image. Slowly he would turn the screw until the star on the right was now sitting on top of the star on the left. A scale showed how far he had moved the lower lens. The scale was engraved with lines so close together he had to read the distance using a microscope.

In the 1830s there was a fierce competition between astronomers to be the first to measure a stellar parallax accurately. Other astronomers concentrated on the brighter stars, believing those to be closer. Bessel, however, looked for a star with a large proper motion. He found it in a dim star in the constellation of Cygnus the Swan. In fact, the star was so dim it did not have a name. Instead, it had a number, 61 Cygni. It had the largest proper motion then known.

In 1838, Bessel announced that 61 Cygni was 632,397 AUs from the sun, or 58,780,000,000,000 miles. This meant that one of the closest stars was still an immense distance away. The distance was so great a new way of measuring distance was needed. Astronomers began speaking of "light-years," the distance light travels in a year.

A light-year is a measure of distance, not time. Light travels about 186,282 miles per second, and a year is 31,536,000 seconds. In a year, light travels about six trillion miles. The modern value for the distance to 61 Cygni is 11.4 light-years.

These two bright stars are (left) Alpha Centauri and (right) Beta Centauri. The circle is where Proxima Centauri is found – which is the closest of the three to Earth.

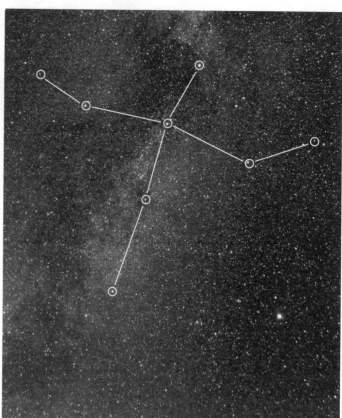

The constellation of Cygnus the Swan was listed by Ptolemy in the 2nd Century AD. It is easily seen during summer and fall in the north, and it contains the grouping of stars known as the Northern Cross.

Since then, closer stars have been found. The sun is the closest, followed by three stars in the constellation of Centauri in the Southern Hemisphere. The closest of the three is Proxima Centauri at about 4.2 light-years. Next is Bernard's star, known as Bernard's runaway star because it has the greatest proper motion of any star. It is too dim to see with the unaided eyes and is about 6 light-years distant. Sirius is at 8.6 light-years, and Procyon is at 11.4 light-years, about the same distance as 61 Cygni.

In 1543, Copernicus was asked why parallax could not be seen if the earth moved. He answered that the stars were too far away. Three hundred years later astronomers succeeded in detecting parallax and measuring the distance. Copernicus had been correct about stars being at a great distance, but he had not realized that stars were so very, very far away and the universe so vast.

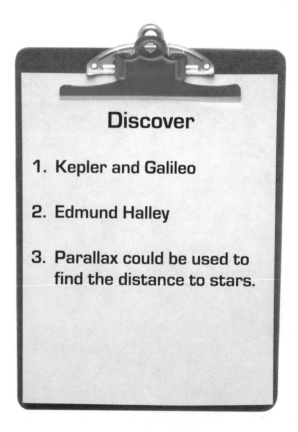

Discover

1. Kepler and Galileo

2. Edmund Halley

3. Parallax could be used to find the distance to stars.

A B C D   1. The one who said, "I have stood on the shoulders of giants" was
(A. Tycho Brahe B. Johannes Kepler C. Galileo D. Isaac Newton).

T F   2. Tycho's success was due to a telescope made for him by Johannes Kepler.

A B   3. Kepler tried to calculate the orbit of (A. Mars B. Venus).

T F   4. Kepler's first law describes the shape of a planet's orbit.

A B C D   5. The scientist who studied motions of objects on earth was
(A. Kepler B. Galileo C. Halley D. Herschel).

T F   6. Gravity grows weaker by the square of the distance.

A B   7. The question about gravity came up when Edmund Halley met with Robert Hooke and
(A. Robert Boyle B. Christopher Wren).

A B   8. Edmund Halley went to the Southern Hemisphere to (A. find a place for a southern
observatory B. precisely locate the position of a network of stars).

A B C D   9. While on St. Helena Island, Halley observed two patches of stars named after the explorer
(A. Christopher Columbus B. Eric the Red C. Henry the Navigator
D. Ferdinand Magellan).

_____ 10. During Edmund Halley's time the most mysterious objects in the solar system were _____.

T F   11. Edmund Halley proved that since the time of the ancient Greeks, the star Polaris had
moved the width of the full moon.

A B C D   12. The first person to measure the parallax of a star was
(A. Fredrich Bessel B. Joseph Fraunhofer C. William Herschel D. Edmund Halley).

_____ 13. Other than the sun, the closest star is Proxima _____.

EXPLORE MORE

Research the life of Edmund Halley and write a two-page report that summarizes his main astronomical achievements.

What did Tycho Brahe do that helped Johannes Kepler? What did Johannes Kepler do that helped Isaac Newton? What did Edmund Halley do to help Isaac Newton publish *Principia Mathematica*? What did Joseph von Fraunhofer do that helped Friedrich Bessel measure the parallax of 61 Cygni?

How does the mass of an object differ from its weight? What is inertia? How do astronauts on the International Space Station "weigh" themselves?

Read about Galileo's experiments with motions on earth. How did he manage to time events when he did not have an accurate clock? How did he reduce the friction of balls that he rolled down inclined ramps? Did he drop two balls of different sizes from the Leaning Tower of Pisa? What is the principle of the pendulum that Galileo discovered? Who used that principle to make the first pendulum clock?

Deep Sky Wonders

One of the best introductions to the stars of the night sky is the Big Dipper. The Big Dipper is by far the best-known star group. It neither rises nor sets. Instead, it circles around and around the north celestial pole, always above the horizon for most people in the Northern Hemisphere. Even as far south as Florida, you can see part of the Big Dipper at any time during the year on any clear night.

The four stars of the bowl and three stars of the handle make the Big Dipper easy to spot. All of the stars except one are of the second magnitude and the remaining one is slightly dimmer. It is a rewarding object for stargazing with the unaided eyes. Look at the star in the crook of the handle of the Dipper. At first you notice only one star. But if your eyesight is good, you will see a smaller star almost in contact with it.

Arab stargazers named the brighter of the two stars Mizar and the dimmer Alcor. Those names, still used today, mean "horse and rider." Natives of North America knew

Explore

1. Why did a comet hunter make a list of 110 deep sky wonders?

2. What is a Coal Sack?

3. What is the Local Group?

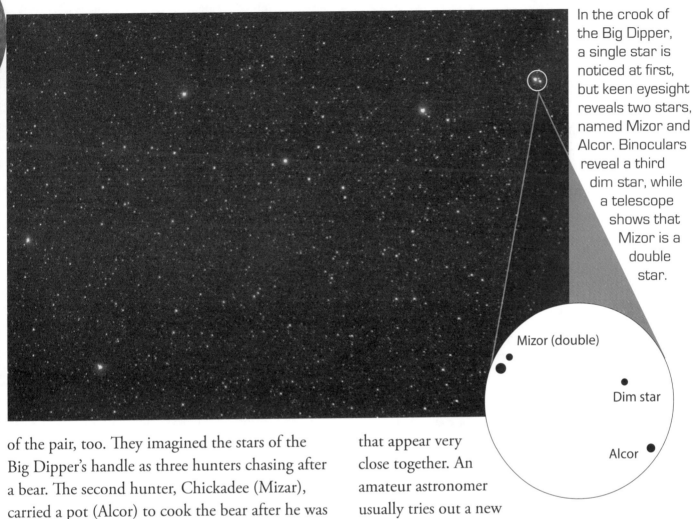

In the crook of the Big Dipper, a single star is noticed at first, but keen eyesight reveals two stars, named Mizor and Alcor. Binoculars reveal a third dim star, while a telescope shows that Mizor is a double star.

Mizor (double)

Dim star

Alcor

of the pair, too. They imagined the stars of the Big Dipper's handle as three hunters chasing after a bear. The second hunter, Chickadee (Mizar), carried a pot (Alcor) to cook the bear after he was caught. But the bear always escaped by circling the northern heavens night after night.

For thousands of years, another secret lay hidden in the twinkle of light from the two stars. The first small telescopes showed that between Mizar and Alcor, and off to one side, is a third dim star. It is visible today in binoculars.

Another secret is hidden in the crook of the Big Dipper's handle. In binoculars or small telescope, Mizar looks notched or egg-shaped. The reason is simple. Mizar is a double star — two stars that appear very close together. An amateur astronomer usually tries out a new telescope by turning it to Mizar. A good telescope easily shows it as a double star.

Where one star is noticed at first, keen eyesight discloses two, while binoculars show three, and a telescope reveals four. Mizar and Alcor prove the full richness of the night sky.

Mizar and Alcor are 11 minutes of arc apart. Astronomers measure the angular size of an object in the sky in degrees, minutes, and seconds.

Moving through a complete circle is 360 degrees.

From horizon to directly overhead is 90 degrees.

The sun and the moon are each about one-half of a degree in angular extent.

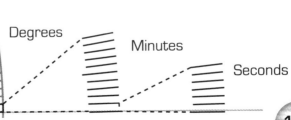

Degrees

Minutes

Seconds

As objects become smaller and smaller, and stars becomes closer together, a degree is too large of a measure. Instead, minutes can be used. A minute of arc is not the same as a minute of time, but they are related to a degree and hour in the same way. Sixty minutes make an hour, and 60 minutes of arc make a degree. The moon is about 30 minutes of arc in diameter. Mizar and Alcor are separated by about one-third the width of the full moon: 30/3 = 10, and Mizar and Alcor are 11 minutes apart.

Many space objects are so small that even a minute of an arc is too large of an angular measure. An even smaller unit is needed to conveniently show the apparent size of the smaller objects. Seconds of arc are used. As you would think, 60 seconds make a minute. Mizar is itself a double star. It and its close companion are 14 seconds of arc apart. Knowing the separation gives an idea of how much magnification is needed to reveal them. A telescope at about 50 power would separate the two stars that make Mizar so they appear about as far apart as Mizar and Alcor: 50 x 14 seconds = 700 seconds, and Mizar and Alcor are 11 minute, or 660 seconds (11 x 60 = 660), apart.

Two stars such as Mizar and Alcor that can be glimpsed as double stars with the eyes alone are called visual double stars. If a telescope is needed to resolve them, then they are telescopic double stars. Mizar and Alcor are visual double stars, while the two stars that make Mizar are telescopic double stars.

Some star pairs are along the same line of sight as seen from earth, but they are far away from one another. In that case, they are optical double stars — they merely look close together. If the members of a double star are bound together gravitationally and orbit one another, then they are binary double stars.

What astronomers have found surprising is that so many stars are true binary double stars. Thousands are now known. One beautiful

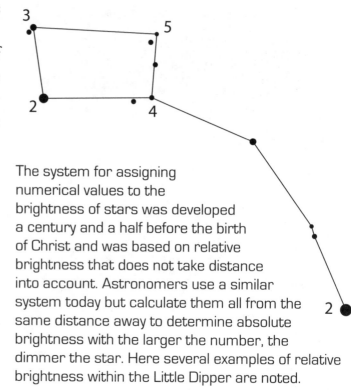

The system for assigning numerical values to the brightness of stars was developed a century and a half before the birth of Christ and was based on relative brightness that does not take distance into account. Astronomers use a similar system today but calculate them all from the same distance away to determine absolute brightness with the larger the number, the dimmer the star. Here several examples of relative brightness within the Little Dipper are noted.

pair is found in Cygnus the Swan. It is a constellation of summer. Stars in this constellation make a star group called the Northern Cross. Deneb, a first-magnitude star, is at the top of the cross. Albireo is a 3rd magnitude star at the base of the cross. Albireo is made of a bright, yellow star and a contrasting blue companion. They are separated by 35 seconds of arc, so a telescope at 40x magnification will reveal them.

The stars of the Big Dipper, along with some nearby dimmer stars, made the constellation *Ursa Major*, which means Great Bear. One of the first mentions of the Great Bear is found in the Bible in the Book of Job. The Lord asks, "Can you bring forth the constellations in their seasons or lead out the Bear with its cubs?" (Job 38:32; NIV). The Hebrew shepherds imagined that the bowl of the Big Dipper formed a bear. The three stars of the handle were three cubs following along behind the Great Bear.

The two outer stars in the bowl of the Big Dipper, known as the Pointers, made a line aimed almost directly to *Polaris*, Latin for pole star. Notice the distance between the two pointer stars and their brightness. Extend the distance between them five times along the line until you arrive at another star of the same brightness as the Pointers. You have found Polaris. It is less than one degree from the north celestial pole. For most purposes, it is more accurate than a compass as a guide to true north.

Polaris is at the end of the handle of the Little Dipper. The stars of the Little Dipper are part of the constellation *Ursa Minor*, which means Little Bear. The Big and Little Dippers are not constellations. They are star groupings called asterisms. They are members of the constellations of Ursa Major and Ursa Minor. Both the Big and Little Bears are circumpolar constellations. They circle around the north celestial pole and remain visible all night long and all year long, provided the night is clear, of course.

The bowls of the Little Dipper and Big Dipper face one another. The Little Dipper is both smaller than the Big Dipper and made of dimmer stars. The Big Dipper is made almost entirely of stars of the second magnitude. The Little Dipper, on the other hand, is made of stars that range from second magnitude down to a dim fifth magnitude star.

Astronomers assign a number that gives a precise value to a star's apparent brightness. The number is its magnitude.

Hipparchus, an ancient Greek astronomer, introduced the system of magnitudes. He lived about 150 years before Christ. Hipparchus organized stars into six groups based on brightness. The brightest he called first magnitude. Those just barely visible to the unaided eyes in a clear, dark sky he called sixth magnitude. In between were groups representing brightness of second, third, fourth, and fifth magnitude. The *larger* the number, the *dimmer* the star. The four stars that make the bowl of the Little Dipper have magnitude of 2, 3, 4, and 5.

Hipparchus made his judgments about brightness using his eyes and comparing one star to another. Today, astronomers are more precise. They set magnitude 1.0 to the average magnitude of the 20 brightest stars. The difference in brightness of a first-magnitude star compared to a sixth-magnitude star is more than a range going from one to six. A sixth-magnitude star is 100 times dimmer than a first-magnitude star. Each magnitude change works out to be about 2.512 times brighter than the next dimmest magnitude. A fifth-magnitude star is 2.512 times brighter than a sixth-magnitude star. A fourth-magnitude star is 2.512 times brighter than a fifth-magnitude star and about six times brighter than a sixth-magnitude star: 2.512 x 2.512 = 6.3.

The brightest star in the bowl of the Little Dipper, magnitude 2, is about 16 times brighter than the dimmest one, magnitude 5. Their

This is the Sombrero Galaxy which was discovered in 1767. The distinctive dark edge is a dust lane, or band, shown against a brighter object as revealed in this image.

difference in magnitudes is 3 (5 − 2 = 3), and 2.512 x 2.512 x 2.512 = 15.85, or about 16.

The brightness of a star as seen from earth is relative brightness. A star may look bright because it is of ordinary brightness but closer to us, or it may look dim but is exceptionally bright but much farther away. The actual brightness of a star is called its absolute magnitude. Absolute magnitude is a measure of the luminosity of a star.

To assign absolute brightness, astronomers calculate how bright stars would appear if they were at the same distance. They use a distance of about 32 light-years. Rigel is a bright, blue star in the constellation of Orion. It is a first-magnitude star, and the sixth-brightest star in the sky. But Rigel is an immense distance away, about 772 light-years. If moved to the standard distance of 32 light-years, it would have an absolute magnitude of −7.84. *Negative* magnitudes are brighter than *positive* magnitudes. Our sun, if moved to 32 light-years away, would appear as a star of about the fifth-magnitude. It would be as dim as the dimmest star in the bowl of the Little Dipper.

Polaris is 434 light-years away. If it were moved to only 32 light-years, then it would appear as an exceptionally bright star of magnitude −3.64.

In addition to double stars, astronomers have also found star clusters. The best-known star cluster is the Pleiades, on the shoulder of Taurus the Bull. Taurus is a large and easily found constellation of winter. A constellation of winter is one that is well placed for viewing at 9:00 p.m. during the winter months of December, January, and February. Stars move from east to west as the earth rotates, so later in the night Taurus sets in the west while stars of spring rise in the east. Aldebaran is the brightest star in the constellation of Taurus. It is a red star of the first magnitude and makes the eye of the bull. Given the red eye, it is easy to imagine that Taurus is an angry bull. A distinctive V of stars makes the bull's horn.

The Beehive Cluster is an open cluster within the constellation of Cancer. The stars are so widely separated and so near the earth that they are best seen with the unaided eyes or with binoculars.

The Pleiades make a spot on the bull's shoulder. The Pleiades has been known since ancient times, and its stars are also known as the Seven Sisters. The dim, seventh star is a test of good eyesight. The night must be absolutely clear, and moonless, and your eyes must be completely adapted to the dark. When trying to glimpse a faint star, look slightly to one side. This puts a more sensitive part of your eye into use. The seventh star will peek in and out as you look at it from the corner of your eyes. If you look directly at the dim star — presto! — it is gone. Binoculars show about 30 stars in the Pleiades. A small telescope at low power will reveal two hundred.

Many of the brightest open clusters were first catalogued by Charles Messier, a French comet hunter. He lived in the late 1700s and early 1800s at about the same time as William Herschel. He came across so many objects that could be confused with comets that he made a list of them. In all, he found 110 of the objects.

Although Charles Messier discovered 21 comets, he is better known today for his list of interesting sky objects. He used a four-inch telescope, which is a popular size for amateur astronomers today. The Messier objects are bright enough to be seen in an ordinary telescope and popular targets for amateur astronomers. By searching them out, amateurs are introduced to five different types of sky objects: open star clusters, globular clusters, external galaxies, glowing clouds of gas, and planetary nebula. More about each type later.

The Beehive open cluster is number 44 on Messier's list. The Beehive is faintly visible as a misty patch to the unaided eyes in the constellation of Cancer the Crab. The ancient Greek astronomer Ptolemy noted it, and it was one of the first objects that Galileo studied with his telescope. Binoculars show it best because it is about one degree across, and most telescopes don't offer a wide enough view to show it in context with the other stars around it.

The Pleiades and Beehive are called open clusters. Many others exist. Open star clusters have a few hundred stars widely separated from one another, but loosely held together by their mutual gravity. They travel through space together. However, they are not quite held strongly enough to stay together. Astronomers believe open clusters are slowly coming apart.

The Coalsack Nebula is a large dark patch seen against the Milky Way from Earth's southern hemisphere.

Globular clusters, however, are more nearly permanent features. The word *globular* means shaped like a globe. Globular clusters have thousands of stars in a very tight arrangement. Their mutual gravity is strong because of their numbers and nearness to one another.

The best-known globular cluster in the Northern Hemisphere is M13 (number 13 on Messier's list) in Hercules, a constellation of summer. Four stars, called the Keystone, make up the main body of Hercules. The globular cluster M13 is found along a line connecting the two stars of the eastern side of the Keystone. M13 is about one-half the apparent size of the full moon. It is an easy object for binoculars. It can be glimpsed with the unaided eye once located with binoculars. A telescope at about 80x begins to show individual stars.

The globular cluster with the greatest relative brightness is Omega Centauri. It is one of the few globular clusters visible to the unaided eyes, and it is about as large as the full moon, but much dimmer. Edmund Halley first described it in 1677 when he went to the Southern Hemisphere to plot the location of a network of stars. In the

The globular cluster within Hercules

1830s, John Herschel, son of William Herschel, recognized it as a star cluster. Today, astronomers believe it is the largest and brightest globular cluster. It has several million stars. The stars near its center are so crowded that they average only about one-tenth of a light-year from one another. If the earth were at the center of the Omega Centauri globular cluster, the light of a thousand suns would dazzle our eyes and nightfall would never occur.

Globular clusters helped reveal the size and shape of our Milky Way galaxy. The Milky Way is a faint, luminous band across the sky. It is brightest in the summer sky, and on moonless nights can be seen as a misty stream running along Cygnus the Swan, between Lyra the Harp and Aquila the Swan and south into Sagittarius the Archer. In Sagittarius, the vast collection of stars shine especially bright.

William Herschel made star counts along the Milky Way and found about the same number of stars in each direction. He used his counts as evidence that the sun was located near the center of the Milky Way. But his conclusion proved incorrect. He was not aware that great clouds of dark gas hid stars in the direction of the Milky Way's center. His star counts failed to detect the region of the sky that had the most stars. They were blocked by gas and dust. The gas is mostly hydrogen and helium. The dust, which is microscopic in size, is carbon and silicon compounds.

Scanning along the Milky Way with binoculars, you can spot dark blots that look like holes in space. They are caused by dark nebula — a collection of gas and dust that is not illuminated by a star. One of these, called Coal Sack, is in Cygnus the Swan. Cygnus lies in one of the parts of the Milky Way that is rich in stars. Along the upper part of the constellation is a strange dark gap in the galaxy. It looks like a hole in space. Although it is not completely free of stars, the Coal Sack's blackness is noticeable in the overall brilliance of the Milky Way.

Astronomers thought the Coal Sack was a hole in space and a window into deeper space. They were mistaken. Instead, a cloud of gas and dust blocked light of the stars that lie beyond it. In other places, invisible clouds of gas and dust reduced the number of stars that can be seen. William Herschel's star counts proved to be in error. His conclusion that the sun was near the center of the Milky Way proved wrong, too.

But by 1900, most astronomers had accepted Herschel's conclusion and put the sun near the center of the Milky Way galaxy. The astronomer who changed their minds was Harlow Shapley.

Harlow Shapley had grown up on a farm in Missouri but dropped out of school with a fifth-grade education. He became a crime reporter for a newspaper. He continued his studies at home. He attended high school and completed the seven years of study he had missed in two years. With his newspaper money, he entered the University of Missouri to study journalism. Unfortunately, the journalism program was delayed a year, so he had to choose another subject. He selected astronomy. He graduated from the University of Missouri in 1910, and then attended Princeton in New Jersey. After earning an advanced degree, he moved to

The Andromeda Galaxy, within the constellation of Andromeda, is located 2.5 million light-years away. This spiral galaxy is one of the most distant objects of the night sky that can be seen with the unaided eye.

California and began working as an astronomer at Mt. Wilson.

About 100 globular clusters were known in 1915, and about 50 more have been found since then. They were not evenly distributed. Most were found in the direction of the constellation of Sagittarius the Archer. Shapley calculated the distance of the globular clusters. He used a recently discovered method using variable stars — stars that changed in brightness. A certain type of variable star revealed its true brightness by the length of time it took to brighten, dim, and return to full brightness again.

Seeing the variable stars at the great distance of the globular clusters, and timing their periods, took the light-gathering power and resolution of the 100-inch telescope. Shapley worked at the problem for five years. In 1920, he found that the globular clusters were positioned above and below the lane of dust clouds in Sagittarius. They were on average about 50,000 light-years away. They hovered over the center of the galaxy, he believed.

Against a lot of opposition, he convinced doubting astronomers that our Milky Way galaxy was larger than previously believed. He also showed that the center lay far away in the direction of Sagittarius. Just as Copernicus moved the earth from the center of the planetary system, Shapley moved the sun from the center of our galaxy.

We see the Milky Way edge on and cannot step away from it to see what it looks like from above. Knowing what the Milky Way galaxy looks like was a problem because the sun is part of it. It is like trying to find the size and shape of a forest (our galaxy) but having our view blocked by the trees (stars).

However, astronomers realized that the Milky Way must be similar to other spiral galaxies that can be seen. One example is the Great Andromeda galaxy (M31). The spiral galaxy is found in Andromeda, a constellation of Fall. It is a little

Ultraviolet photograph reveals details of the Triangulum (or Pinwheel) Galaxy.

brighter than magnitude 4, so it is readily visible with the eyes alone. On a clear, moonless night free of haze, the galaxy can be traced out with the unaided eyes for three degrees — about six times the diameter of the full moon. It covers the field of view in binoculars. A telescope can only show the central nucleus. Astronomers have calculated its distance as 2.5 million light-years. It is the most distant object that can be seen with the eyes alone.

The true shape of the great spiral galaxies like M31 were first revealed in the sketches that Lord Rosse made with the Leviathan, his 60-inch metal reflector telescope. Time-lapse photographs show much more. M31 has several small, irregular companion galaxies, and the two brighter ones are similar to the Milky Way's Large and Small Magellanic clouds. It, too, has a crown of globular clusters around its center, and dark lanes of gas and dust.

M33 in Triangulum is another well-known spiral galaxy. The Triangulum constellation is made of three stars and is located southwest of Andromeda. M33 is nearby as galaxies go, being the next-door neighbor to the Andromeda spiral and our own Milky Way. M33 is slightly brighter than magnitude 6, but the light is spread out over an area twice that of the full moon. Binoculars are needed to show it. It is a spectacular sight under the right conditions: the night must be absolutely clear, moonless, and the observer's eyes must be completely dark adapted. M33 appears as a pinwheel, and an optical illusion sometimes makes it appear to be on the near side of the background of stars, hanging only a few hundred feet overhead.

We see M31 and M33 more or less broadside. However, other galaxies appear edge-on. A dark lane obscures their central disk, in the same way the central disk of the Milky Way is hidden. These observations have convinced astronomers that the stars of the Milky Way form a spiral with arms similar to M31 and M33.

Our Milky Way galaxy is a spiral about 110,000 light-years across containing about 300 billion stars. The central hub is about 16,000 light-years thick. The sun is midway out from the center in one of the spiral arms. At that location, the arm is about 3,000 light-years thick. The average distance between stars at the center of our galaxy is 1.3 light-years, but in our neighborhood the average distance is 4 light-years.

The Milky Way galaxy is not the only island universe. It is but a single, typical galaxy among many millions of such systems that have been photographed with large telescopes. Galaxies show a tendency to occur in clumps or clusters. For instance, within a region of five million light-years, astronomers have found three major galaxies (Milky Way, Andromeda galaxy, spiral galaxy in Triangulum), several irregularly shaped galaxies like the Magellanic Clouds, and many miniature collections of stars known as dwarf galaxies. They form a galactic cluster called the Local Group.

The Local Group is not the end of the universe but only a small neighborhood embedded in a much larger universe. The Hubble Space Telescope has taken a photograph that reveals thousands of galaxies in a region billions of light-years across. The size of the universe is so vast as to be beyond comprehension. However, astronomers continue to probe as deep into space as they can. Some astronomers believe there is an actual edge or end of the universe. However, they also agree that no telescope or other instrument that we currently possess can see deep enough into space to image the edge of the universe.

Gas and dust clouds hid the true nature of the Milky Way. But not all gas and dust is dark. Nearby stars can make the gas glow. Instead of a dark nebula, the gas forms a bright or emission nebula. The word *emission* means it emits light.

One of the best-known emission nebulas is the Orion Nebula. Orion is a winter constellation and easy to recognize. He has a bright red star, Betelgeuse on his left shoulder (northeast corner) and a bright blue star, Rigel, on his right foot (southwest corner). He also has stars that make the belt and sword. The Great Nebula of Orion (M42) appears to the unaided eye as a misty star in the middle of the sword of Orion. Binoculars enhance the view. A telescope at 200x power reveals the central group of double and multiple stars that cause the gas to glow. Time exposure photographs with professional telescopes reveal a vast glowing field of gas with colors of red and blue.

Planetary nebulas are another type of glowing nebulas. They have nothing to do with planets. To early telescope users they could be mistaken for planets because they have a round shape. They have a shell of expanding gas around them. The star became unstable and blew off some of its material before settling back down. The gas expands outward and will eventually fade away.

One of the best-known planetary nebulas is the Ring Nebula in the constellation of Lyra the Harp. A dim star in the center causes the hydrogen gas around it to glow. But it is dim and the ring is

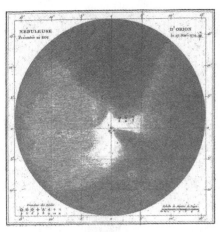

Messier's drawing of the Orion Nebula in *Mémoires de l'Académie Royale* (left); the nebula as seen today (right)

An 1883 photograph of the Orion Nebula with long exposure showing stars and nebulas invisible to the human eye (above); at right, a much more detailed modern example

barely visible in a 4-inch telescope at 120x. The Dumbbell planetary nebula (M27) is in Vulpecula. The name means little fox. Of the 88 constellations, Vulpecula is one of the faintest. Its brightest star is 4th magnitude. The Dumbbell is easy to see. It is twice as bright and 50 times as large as the Ring Planetary. A three-inch telescope gives vague details at 80x. It may be detected in binoculars once located.

A planetary nebula is around a star that exploded but the star survived. The Crab Nebula (M1) in Taurus is another story. It received its name from Lord Rosse, who sketched its glowing filaments of light as a crab. It is what remains of the supernova recorded by Chinese astronomers in A.D. 1054. They called it a "guest star," and it was visible in daytime. Europe was deep in what became known as the Dark Ages. At that time, Europeans had very little interest in astronomy. So if they saw the supernova, they must have ignored

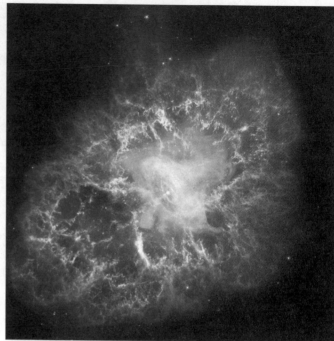

The Ring Nebula (left) was discovered in 1779 and remains one of the more recognizable nebula for those exploring the wonders of the night sky. The Crab Nebula (right) was identified in 1731 – and is known for the radiation that is emitted from it, the remnant of a supernova.

it. No European records of its appearance have been found.

The Crab Nebula is located where the Chinese astronomers saw the bright, new star. Photographs taken several years apart showed that the Crab Nebula was expanding. Tracing the expanding gas backward, astronomers realized it must have been a star that exploded in 1054.

Only a small portion of the 110 Messier objects have been described in this chapter. His catalog is only a small listing of the unusual objects that make the night sky so fascinating. Modern astronomers now use a much larger catalog — the New General Catalog — to list objects. It contains more than 7,000 entries.

Astronomers who studied the stars faced a daunting task. They had to study their subjects from afar. Despite this, by the middle of the 1900s they had made remarkable progress. They could gauge a star's distance, identify the chemicals that made it, measure its surface temperature, and

determine its absolute magnitude. They could calculate its direction of travel and clock its speed.

How they made such remarkable progress is explored in the next chapter.

Discover

1. To avoid confusing them with comets.

2. A dark region in the Milky Way caused by gas and dust.

3. A cluster of galaxies.

T F   1. The Big Dipper is easy to see because all seven of its stars are of the first magnitude.

A B   2. The moon is about _____ in angular extent. (A. 30 minutes   B. 30 degrees)

A B   3. The Big Dipper is (A. an asterism   B. a constellation).

A B   4. The Little Dipper is part of the constellation known as (A. Ursa Major   B. Ursa Minor).

A B   5. Double stars bound together gravitationally are called _____ double stars.
(A. binary   B. visual)

A B   6. Both Ursa Major (Big Bear) and Ursa Minor (Little Bear) are _____ constellations.
(A. circumpolar   B. southern)

A B C D   7. The Pleiades are an example of a (A. double cluster   B. globular star cluster
C. galaxy   D. open cluster).

T F   8. A first-magnitude star is six times brighter than a sixth-magnitude star.

T F   9. A star of −3.0 magnitude is brighter than one of 4.0 magnitude.

_____ 10. Charles Messier was a French _____ hunter.

A B C D   11. The largest and brightest globular star cluster is _____ Centauri.
(A. Alpha   B. Beta   C. Omega   D. Proxima)

A B C D   12. William Herschel's star counts were in error because (A. dust and gas blocked more distant
stars   B. of interference by city lights   C. his location was too far north of the equator
D. of a tarnished mirror in his telescope).

A B C D   13. Harlow Shapley found the location of the center of the Milky Way galaxy
(A. by examining the shape of external galaxies   B. by making accurate star counts
C. by plotting the positions of globular clusters   D. using x-ray photography).

EXPLORE MORE

Research the life of Harlow Shapley and write a two-page biography that summarizes his astronomical achievements.

From the ancient Phoenicians to the Pacific Islanders to the great seafaring nations of Spain and Great Britain, mariners developed ways to navigate seas and oceans. Research how they used stars for navigational purposes.

Find monthly star charts of the circumpolar constellations and notice how the position of the Big Dipper changes from month to month. In which month is it directly above the North Pole in a position to pour out its contents? When does the bowl point west? East? When is the bowl low in the north?

What is procession of the equinoxes? What causes it? Who was the first astronomer to discover procession of the equinoxes? How has this changed the location of the North Celestial Pole? Was Polaris always the North Star? What other effects has it had on the location of constellations?

Stars

The spectroscope is the second-most-important instrument astronomers use. The telescope is first, of course. The spectroscope is one of the most powerful tools astronomers have. This instrument can analyze the light of a star and give information about the star's chemical composition, temperature, relative motion with respect to the earth, and how fast the star is rotating. A spectroscope that is designed to take photographs is known as a spectrograph.

A spectroscope spreads the light of a star into the spectrum — the colors of the rainbow: red, orange, yellow, green, blue, and violet. Bright and dark lines cross a star's spectrum. The bright lines are emission lines and are caused by hot elements that emit light. The dark lines are caused by cooler elements that absorb light.

Each chemical element — hydrogen, helium, calcium, oxygen, etc. — has a different pattern of lines. Hydrogen, which is found in all stars, has a set of lines that are especially easy to recognize.

Explore

1. What are main sequence stars?

2. Why did astronomers think the companion of Sirius was a dark star?

3. Why are Cepheid variables called optical yardsticks?

Astronomers collected light from the sun with a telescope and passed it through a spectroscope. The spectroscope revealed the elements in sunlight by the pattern of lines. Astronomers found that the sun was made of the same elements as those found on earth — with one exception.

In 1868, a French astronomer named Pierre Janssen detected lines in the sun's spectrum that did not match any element on earth. Pierre Janssen asked an English astronomer, Joseph Lockyer, to confirm his discovery. Joseph Lockyer agreed that the solar element did not match any known earthly element. He gave the new element the name helium, from the Greek word for "sun." Twenty-five years later, helium was found on earth.

Hydrogen is the most common element in stars. Astronomers keyed on lines in the spectrum made by glowing hydrogen. Those with especially noticeable hydrogen lines were called A-type stars. Those with especially weak hydrogen lines were given the letters farther down the alphabet. M-type stars, for instance, had lines produced by metals but practically none caused by hydrogen.

This arrangement proved not to correctly reflect the temperature of the stars. Hydrogen cannot emit light if the star is so hot that electrons have escaped from hydrogen atoms. Instead, lines from helium are more noticeable. Nor can hydrogen emit lines if the star is so cool that the electrons remain inactive. Instead, lines from metals and molecules of

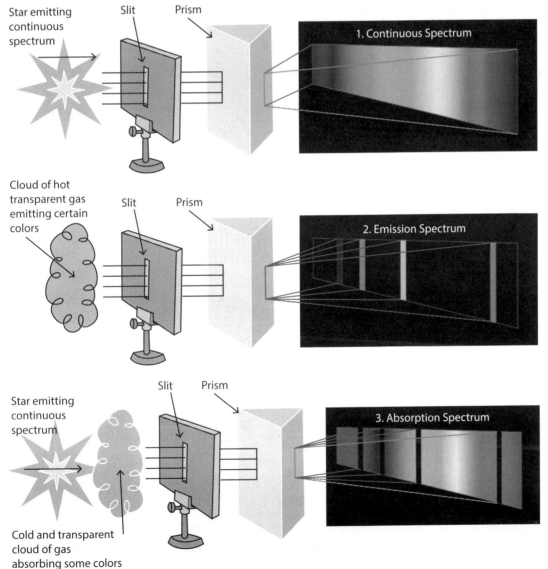

Star emitting continuous spectrum

Slit

Prism

1. Continuous Spectrum

Cloud of hot transparent gas emitting certain colors

Slit

Prism

2. Emission Spectrum

Star emitting continuous spectrum

Slit

Prism

3. Absorption Spectrum

Cold and transparent cloud of gas absorbing some colors

Three ways scientists can determine the composition of distant objects in space based on a spectrum of colors: 1) when nothing is clouding the object like an atmosphere, you have a wide range in the spectrum; 2) if the object has hot gases emitting light, these appear as bands on the spectrum; and 3) you can also calculate objects behind other objects like a gas cloud, based on the absorption within the spectrum.

titanium oxide are more prominent. Electrons in metals require less energy to put them in motion. So metals show up in the spectrum of cooler stars. Hydrogen doesn't show up in the spectrum of very hot stars or in very cool stars.

Astronomers recognized color as an indication of surface temperature. Blue stars have the hottest surface (15,000 to 50,000°F). White stars are not quite as hot (11,000 to 15,000°F). Yellow and orange are in the middle range (5,000 to 11,000°F). Red stars are much cooler (less than 5,000°F). The sun has a surface temperature of about 10,000°F. The sun is a yellow star.

Based on how certain spectrum lines grew weaker and others grew stronger, astronomers reordered the letters based on the temperature of the stars. Starting with the hottest stars and moving toward the cooler ones, the classes are: B, A, F, G, K, and M. Astronomers remember the sequence by the mnemonic "Be a fine girl, kiss me." (Or, in case of a female astronomer: "Be a fine guy, kiss me.") Four other spectral classes are O, R, N, and S. But these are very rare types.

B and A stars shine with an intense blue light. They radiate great amounts of heat and light. Spica, the brightest star in Virgo the Virgin, is a B-type star. Spica's surface temperature is almost 33,000°F — very hot indeed. It can be found by following the arc of the handle of the Big Dipper to the star Arcturus and continuing the same distance to Spica. Rigel, the bright blue star in Orion, also has a B spectral classification.

F, G, and K stars are white, yellow, or orange in color. Proceyon in Canis Minor has a surface temperature of 11,000°F, making it an F-type star. Our sun is an example of a G type star. Its surface temperature is about 10,000°F, which is more or less in the middle of the temperature range. Pollux, one of the main stars in Gemini the Twins, is a cooler K-type star.

Dominant Color and Temperature of Stars			
Class	Temperature	Color	Example
B	10,000 -30,000	Blue	Rigel, Spica
A	7,500 - 10,000	Blue	Vega, Sirus
F	6,000 - 7,500	White	Proceyon
G	5,000 - 6,000	Yellow	the Sun
K	3,500 - 5,000	Orange	Pollux
M	2,000 - 3,000	Red	Antares, Betelgeuse

Stars of class M are cooler and vary in color from orange to dull red.

For most stars, brightness and temperature are directly related. Hot stars like Spica are luminous. Cooler stars such as Pollux are not as luminous. For instance, Spica has an absolute magnitude of -3.55 while Pollux has an absolute magnitude of 1.08. At the same distance from earth, a hot, blue star would be brighter than a cool, red star. Stars whose temperature and magnitude are in agreement are called main sequence stars.

Any star that has a high temperature but is cool, or one that has a low temperature but is bright, is not a member of the main sequence. Both Antares and Betelgeuse are very luminous despite having a low surface temperature. Their surface temperature makes them of type M (a cooler star). However, their large size makes them very luminous. They are a type of star known as Red Giants, which are cool but luminous due to their large size. At the other end of the scale, White Dwarfs are hot but dim due to their small size.

A casual survey of the night sky might show that the hotter stars are more common. But that is not the case. They are hotter, brighter, and easier to see because of their absolute magnitude. We can see bright stars for hundreds of light-years in all directions. Cool, dim stars are harder to see. Fewer of them are visible. Not a single main sequence M-type star can be seen with the unaided eyes.

(Red Giants are M-type stars and are visible to the unaided eyes, but Red Giants are not main sequence stars.)

Telescopic star counts show that cool stars of low luminosity outnumber hot stars of high luminosity. In a sample of 1,000 stars, 734 would be stars of the cool, dim class M, and about 260 of the brighter, main sequence types. The remaining stars are special cases that are not members of the main sequence — Red Giants and White Dwarfs.

The size of stars cannot be measured directly. They are too far away for telescopes to magnify them large enough to measure. But Albert Michelson, American inventor and scientist, found a way to use light to reveal information about the size of a star. He was born in Germany, but when he was four his parents came to America. He graduated from the Naval Academy in 1873. He taught science there, but his passion was learning more about light.

Light is made of waves, and waves can interfere with one another. When a pebble is tossed into a pond, the splash creates waves that radiate outward. A wave is made of a crest and a trough. If a second pebble is tossed into the water, it too causes waves.

When the waves from the two pebbles meet, they will affect, or interfere, with one another. If they meet crest to crest, then they produce a higher crest. But if a crest meets a trough, they will cancel one another. In the location where they cross, the water will be undisturbed for a moment.

Light is a wave motion, too. It is made of electromagnetic waves. Most waves need a medium in which to travel. Sound travels in air, and waves on a pond travel in water. But light waves need no medium to carry them. They can travel in a vacuum. Light waves can interfere with one another as water waves do. Where light waves meet crest to crest, a bright fringe is seen. Where light waves meet crest to trough, a dark fringe is seen.

Light waves are incredibly tiny. It would take 2,000 light waves to be one millimeter long. Special equipment is needed to see the bright and dark fringes caused by light interference. Albert Michelson became an expert at making equipment to reveal the interference of light waves.

Albert Michelson realized that light from one edge of a star that comes straight into the tube of a telescope has a slightly shorter path than one that comes from the other edge of the star. No matter

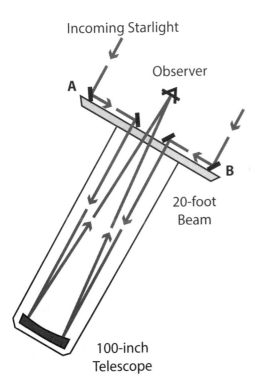

Incoming Starlight

Observer

A

B

20-foot
Beam

100-inch
Telescope

An interferometer helps to make telescopes more accurate in how they handle the interference of light waves. It basically adjusts how the waves are received to mimic two separates telescopes separated from one another.

how long and skinny a right triangle is, the longest leg is always shorter than the hypotenuse. He built an instrument called an interferometer, that measured the difference in distance the two waves traveled using the interference of the two streams of light waves.

Rather than being at the eyepiece, the Michelson interferometer sat on the top of the tube of the 100-inch Hooker telescope at Mt. Wilson. The 100-inch (8.3 feet) was the largest telescope in the world at the time. The interferometer was a metal beam 20 feet wide with two mirrors set at a 45-degree angle at each end. The mirrors directed light to another pair of mirrors that reflected light down the tube to the eyepiece. The mirrors made a light path like a periscope set on its side.

Each mirror received light from a different part of the star — from the left edge of the star and the right edge of the star. The mirrors at the top of the tube could be moved, and in doing so, bright and dark interference fringes would be seen in the telescope.

It was very delicate work. The astronomer had to judge when the fringes were at the brightest and dimmest. The first star Michelson measured was Betelgeuse. It proved to have an apparent diameter of 0.045 seconds of arc. Although a very tiny angle, when compared to the distance to Betelgeuse, it showed the star to have an enormous diameter. Betelgeuse was far bigger than the sun. If Betelgeuse replaced the sun, Mars would be skimming along its surface.

In all, Michelson managed to measure the diameters of seven stars: Betelgeuse, Arcturus, Antares, Aldebaran, a star in Hercules, one in Pegasus, and Mira, a star that varied in brightness. All of these were found to be exceptionally large stars. They came to be called Red Giants.

To measure the diameters of smaller stars would take baselines longer than 20 feet. But the dome for the 100-inch telescope could not be opened

Sometimes smaller or dimmer stars can be lost in the brightness of another object. Sirius A is the brightest of the night stars – so it took an overexposed image to reveal its tiny neighbor, Sirius B nearby. Relatively close at only 8.6 light years away, the two revolve around each other every 50 years.

any wider. Telescopes today are often built in pairs separated from one another. Keck I and Keck II telescopes on Mount Mauna Kea in Hawaii are examples. They work together and are connected by fiber optics and computer systems. The pair of telescopes can make measurements that would be impossible with a single telescope working alone.

At the other end of the scale from large Red Giants are small White Dwarfs. Telescopes easily reveal large, bright stars such as Red Giants. But they are rare. Only about one star in a hundred is a Red Giant. White Dwarfs, on the other hand, are eight times more common, but because of their dimness, they remained hidden until the 1840s.

One such tiny star was detected because of its effect on Sirius, the Dog Star. Sirius has a proper motion across the sky. Edmund Halley noticed that it had moved in the 150 years since Tycho plotted its position. After measuring the parallax of 61 Cygni, Friedrich Bessel continued to precisely measure the positions of other stars, including Sirius. Bessel noticed that Sirius did not follow a straight-line path. Instead, it traveled so that part of the time it was above the straight line of its motion and part of the time it was below the line. It moved across the sky in a wavy line.

Friedrich Bessel concluded that it had a massive, but unseen, companion star. The companion pulled it off course. The two stars were going around their common center of gravity. However, none of the telescopes in use in the 1840s would show the star traveling through space with Sirius. Because it could not be seen, Bessel and other astronomers decided that it was a burned-out star. They believed that the companion of Sirius was nothing but a dark, burned-out cinder.

Despite being invisible, it had to be big. To pull Sirius off course it would need to be about the same mass as our sun.

Twenty years later, in 1862, the American telescope maker Alvan G. Clark finished a new 18-inch lens he'd made at his telescope shop in Cambridgeport, Massachusetts. It was the largest lens in the world at that time. Alvan G. Clark would later make the 26-inch lens for the Naval Observatory that Asaph Hall used to discover the moons of Mars. Clark also made the 40-inch lens for the Yerkes Observatory in Williams Bay, Wisconsin.

The 18-inch telescope had been ordered by the University of Mississippi but could not be delivered. The Civil War stopped shipments between Massachusetts in the north and Mississippi in the south. The telescope eventually ended up at the University of Chicago.

But before he let a telescope leave his shop, Clark always tested it thoroughly. When he turned the lens on Sirius, he saw what looked like a defect in his lens. Next to Sirius was a sparkle of light. He quickly moved to inspect another bright star and saw no sparkle. Checking Sirius more carefully, he

While telescope design has been refined and improved since they were first invented, the basics of their design and use have stayed remarkably consistent. Professor Asaph Hall using the Naval Observatory's 26-inch telescope in 1924; he would use the telescope to make a number of important discoveries.

realized the sparkle was a tiny star nearly lost in the glare of the light of Sirius.

Alvan Clark had discovered the so-called dark companion of Sirius. It was not dark at all. The main star, Sirius, became known as Sirius A. The dim companion became know as Sirius B.

Once astronomers could see Sirius B, they immediately began making calculations. They know from parallax the distance to Sirius A. From the relative brightness of Sirius B, magnitude 8, they found it to be many times dimmer than our sun.

The spectroscope showed that the surface of Sirius B was especially hot — slightly hotter than the surface of our sun.

Sirius B became a puzzle. It was hotter than our sun, more massive than our sun, but more than 100 times dimmer than our sun. To reconcile all of these observations, it had to be smaller than the sun. Although the surface was hot, its small surface area did not radiate much light. Sirius B had to be a tiny, hot star, but with all of its mass packed into a small space. It had to be only about the size of earth.

Because it was white hot but small, Sirius B became known as a White Dwarf. Many others have been found. Procyon, the brightest star in Canis Minor (Little Dog), also has a White Dwarf companion.

The material that makes a White Dwarf is packed together with incredible density. Its density is about a million times as dense as the material that makes the sun. Material from a White Dwarf

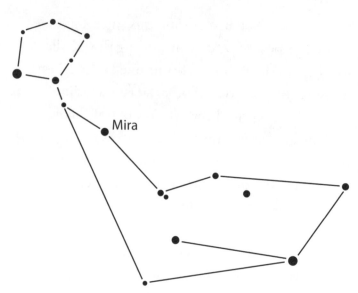

Mira in the Cetus constellation is an example of a variable star that cycles through various magnitudes for almost an entire year at a time between brightest magnitude to brightest magnitude.

the size of a pencil would weigh as much as the Washington Monument.

By the time all of this was worked out, astronomers knew that substances were made of atoms. Most of an atom was empty space. A tight, dense nucleus of protons and neutrons was at the center with electrons orbiting much farther away. The atoms of a White Dwarf had been crushed down, forcing most of the space out of the atom.

Stars can change in brightness. Polaris is a variable star. Over a period of four days, it becomes dimmer by about 10 percent. That is not enough to be easily detected except with special equipment. However, other stars can vary in brightness by several magnitudes. Some, like Polaris, change in brightness because the star itself pulsates. It grows bigger and brighter. Others, change in brightness because a dimmer star orbits them. The dimmer star comes in front of the primary star and blocks some of its light as seen from earth. Stars of this type are called eclipsing binaries.

A star that actually grows dimmer and brighter is called an intrinsic variable. One of the best

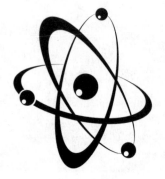

known true variable stars is in the constellation of Cetus the Whale. Cetus is a constellation of Fall, located south of Pisces the Fish. Rather than a whale, Cetus represents a sea monster, and some say it represents the leviathan mentioned in the Bible in Job 41.

In 1596, before the invention of the telescope, a Protestant minister named David Fabricius reported that one of the stars in Cetis varied in brightness. For five months it was invisible to the unaided eyes. Then it increased in brightness to become the brightest star in the constellation. Fabricius lived during the time of Tycho Brahe and Johannes Kepler. Most astronomers in those days believed stars were unchangeable. Stars didn't move and didn't change in brightness. Fabricius's discovery offered one example to prove this idea false.

The variable star in Cetus was later named Mira, which means wonderful. It can brighten to magnitude two or even brighter and then fade to magnitude eight or ten. That magnitude range represents a thousandfold change in brightness.

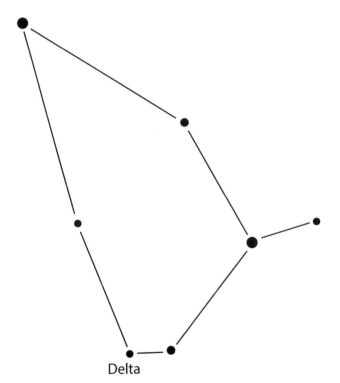

Delta

Delta in the Cepheus constellation is an example of a Cepheid variable star, which keeps an unchanging rate of variability.

The complete cycle from maximum brightness back to maximum brightness takes about 11 months. Because its period is long for a variable star, Mira is known as a long-period variable.

Most long-period variables have periods that range from three months to two years. But they are not regular, and their maximum can occur days or even weeks earlier or later than predicted. A group of amateur astronomers known as the American Association of Variable Star Observers helps professional astronomers track changes in variable stars. Members of the association select stars to watch and record the magnitude. They use nearby stars of known magnitude to compare with the variable. The information is pooled and averaged. They develop a light curve for each variable star. A light curve is a graph showing how the star changes in brightness. Along one axis is the time in hours or days. The other axis is the magnitude.

Mira is a type M star. The surface temperature can vary from about 4,200°F to a low of 3,000°F, which is cool for a star. The interferometer of Albert Michelson's showed that Mira was about 400 times the diameter of the sun. As it pulsates in brightness, it also increases in diameter by about 20 percent. The entire star swells and shrinks. Mira is a Red Giant and not a member of the main sequence.

One type of variable star has proven very useful to astronomers to measure distance. The direct measurement of distance using parallax is good out to about 325 light-years. After that, the parallax is too slight to be measured accurately with today's instruments. However, Cepheid variables can then be used. They are often part of binary stars and star clusters. The brighter ones can even be seen in external galaxies.

Cepheid variables are named after a star in the constellation of Cepheus the King. Cepheus is located near the north celestial pole next to Cassiopeia. In Greek mythology, Cassiopeia was his wife and Andromedia was his daughter. All three constellations are located near one another. Cepheus has a shape like an open envelope with the top star about 12 degrees from Polaris. Face north during the fall to locate this constellation.

The chief attraction in Cepheus is the star Delta Cepheus. Delta is the fourth letter of the Greek alphabet. When stars in a constellation are not named, astronomers give them Greek letters in order of their brightness. Delta Cepheus is the fourth-brightest star in Cepheus.

Delta Cepheus changes in apparent magnitude from 3.7 to 4.9 in 5.37 days. The magnitude change is large enough to be tracked with the unaided eyes, provided its brightness is compared with nearby stars. John Goodricke of England first discovered this variation in 1784. He was 19 years old at the time of his discovery.

The period of Delta Cepheus is very exact. Unlike the long-period variables, this variable star and others like it maintain an unchanging rate of variability. The maximum brightness comes precisely as predicted. After Delta Cepheus's discovery, many others have been found. Stars like Delta Cepheus are known as Cepheid variables. All of them maintain a predictable period, and their change in brightness remains the same, at least over a period of several years.

Cepheid variables have a distinctive light curve, shaped like a shark's fin. The periods of Cepheid variables range from 1 day to as much as 100 days. It would have been useful to know how the absolute luminosity of the Cepheid variables compared to one another. For instance, were those of longer period more luminous or less luminous than those with shorter periods? When the Cepheids were first discovered, astronomers could not answer that question. None of the Cepheids were close enough to have a parallax. Their distances were unknown, so their absolute luminosities could not be calculated.

In 1912, Henrietta Swan Leavitt began studying Cepheid variables in the Small Magellanic Cloud. She worked at the Harvard Observatory but examined photographs taken at an observatory in Peru. From that South American country, the Small Magellanic Cloud was readily visible.

The Magellanic Clouds were external galaxies. All stars including the Cepheids in the Small Cloud were about the same distance from us. The distance from the sun to the clouds was a much larger factor than the distance between the stars inside the cloud.

When Henrietta Leavitt compared the relative brightness of two stars, she knew she was also comparing their absolute luminosity. If one Cepheid in the Small Cloud measured brighter than another, it did so because it was actually more luminous than the other.

As she worked with the Cepheids, she made a remarkable discovery. The brighter the Cepheid variable, the longer its period. A Cepheid in the Small Cloud with a magnitude of 15.5 had a period of two days. One with a magnitude of 14.8 had a period of five days. One with a magnitude of 12.0 had a period of 100 days. (Remember, the smaller the magnitude, the brighter the star.)

Henrietta Leavitt plotted a graph with the period in days along one axis and the magnitude

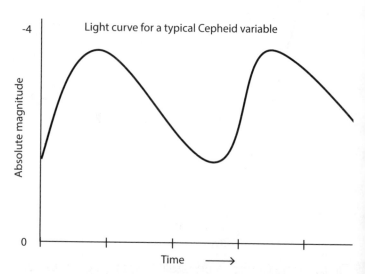

A graph showing the consistency of a Cepheid variable

AB7 within the Magellanic Clouds is an example of a star with very strong stellar winds, ejecting energetic particles, and featuring a surface temperature of 120,000 degrees, making it among the hottest known to astronomers

along the other axis. Then she connected the dots with a smooth line. The line, which gently curved upward, related period to luminosity. Suppose an astronomer wanted to know the magnitude of a Cepheid. All he would need to do was measure the time the star took to change from bright to dim and back to bright. The period in days gave the brightness.

Her discovery, however, was like Kepler's third law of motion. His law let astronomers make a scale model of the solar system using the distance of the earth from the sun as a guide. However, in Kepler's day astronomers did not know the actual distance of the earth to the sun. In the same way, Henrietta Leavitt's light curve could only show relative distances. The actual distance to the Little Cloud was not known.

With the curve, astronomers could compare the relative distance to globular clusters, the relative distance to the center of our galaxy, and the relative distance to the external galaxies, such as the ones in Andromedia and Triangulum.

To find the actual distance, astronomers needed to know the actual distance to at least one Cepheid. In the early 1900s, telescopes could not measure parallax of stars more than 150 light-years away. No Cepheid was inside that distance. The closest one did not show a measurable parallax.

Although parallax was the best way to measure distance, astronomers had other methods. The spectroscope once again helped solve the problem.

Astronomers can use the spectroscope to measure a star's speed. In addition to showing the composition of a star, a spectroscope reveals a Doppler shift due to motion. The shift is named after Christian Doppler, an Austrian physicist who lived in the 1800s. He showed that the frequency (pitch) of sound waves changes if the source of the sound waves is moving. His discovery applied equally well to light waves. The frequency (color) of light waves changes if the source of the light waves is moving.

If a star is moving toward us, the lines in its spectrum are shifted toward the blue end. If a star is moving away from us, the lines in its spectrum are shifted toward the red end. The amount of shift is related to its speed. The greater its speed, the greater the shift. The spectroscope, then, can be used like a radar gun. It can be pointed at a star and tell us how fast the star is going either toward us or away from us.

Stars move in all directions. From earth, that motion is seen in two separate parts. The part of its motion away from or toward us is the radial speed as revealed by the spectroscope. The part of its motion across the line of sight is the transverse speed. It is measured in seconds of arc per year.

To calculate the distance to a star, astronomers needed both the star's transverse motion in seconds of arc and its transverse speed. They could then apply a branch of mathematics known as trigonometry to calculate the actual distance to the star.

They were caught in a problem. Only by knowing the distance could they change seconds of arc into actual units of speed such as miles per second — yet they needed the actual speed to measure the distance. Using a complicated procedure, they substituted the radial speed revealed by the spectroscope for the transverse speed. The method would only give acceptable results if several stars were used and the results averaged together.

Astronomers made calculations of distances for several Cepheid variable stars and averaged the results. With known distances to the Cepheids, the period-luminosity relationship changed from a relative scale to an absolute scale. Astronomers merely had to measure the period of a Cepheid variable. Then they looked up its absolute magnitude and distance using the period-luminosity table.

No longer did the Cepheids give merely a scale model of the Milky Way and its neighbors. Now, actual distances could be calculated. The Small Magellanic Cloud was about 200,000 light-years away. The Large Magellanic Cloud was a little closer at 160,000 light-years away. The great spiral galaxy in Triangulum was about three million light-years distant.

Today, telescopes have improved enough to measure the parallax and distance of about six Cepheids. Results for these Cepheids confirmed that the earlier estimates of distance were correct.

Cepheids have become an optical yardstick to measure distance. They are luminous stars and can be seen from a great distance away. They are visible in nearby galaxies and are used to measure the distance to the members of the Local Group of galaxies.

The vast nebula known as NGC 604 is almost 1,500 light years across – large enough to be viewed by telescopes on Earth, although this image is from the Hubble Telescope. It contains over 200 hot stars, all larger than our sun.

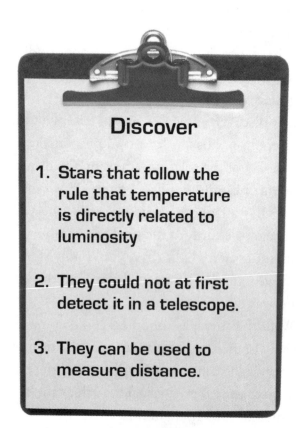

Discover

1. Stars that follow the rule that temperature is directly related to luminosity

2. They could not at first detect it in a telescope.

3. They can be used to measure distance.

_____ 1. The element found on the sun before being found on earth was _____.

_____ 2. The sun has a surface temperature of about _____ °F.

A B   3. B and A type stars shine with an intense blue light and have a surface that is _____ than other stars. (A. hotter B. cooler)

_____ 4. Cool, dull red stars are of type _____.

T F   5. Red Giants and White Dwarfs are examples of main sequence stars.

T F   6. Cool stars of low luminosity outnumber hot stars of high luminosity.

A B C D   7. Finding the size of stars was first done with (A. an interferometer B. a light curve C. the period of a long-term variable D. a spectroscope).

A B   8. The first stars whose sizes were measured were (A. Red Giants B. White Dwarfs).

T F   9. The first person to see the dim companion of Sirius was the American telescope maker Hans Lippershey.

A B   10. A White Dwarf is _____ dense than the sun. (A. more B. less)

A B   11. Mira is an example of _____ star. (A. Cepheid variable B. long-term variable)

A B C D   12. The astronomer who developed light curves for Cepheid variable stars was (A. Frederick Bessel B. Pierre Janssen C. Henrietta Leavitt D. Albert Michelson).

A B   13. If a star is moving away from the earth, its spectral lines are shifted toward the _____ end of the spectrum. (A. blue B. red)

T F   14. Cepheid variable stars are bright enough to be seen in nearby galaxies.

EXPLORE MORE

Research the life of Albert Michelson or Henrietta Swan Leavitt. Write a two-page biography, and summarize their astronomical achievements.

The spectroscope today is made not with a prism but with a diffraction grating. What is diffraction? Who discovered the diffraction of light? How does it separate the colors of light? How are diffraction gratings made? What is its advantage over a glass prism for revealing the emission and absorption lines in a star's spectrum?

What are some of the typically high temperatures on earth: the highest setting on a kitchen oven, the temperature of a kiln for glazing pottery, the temperature of an oxyacetylene welding torch, etc.? What is the highest temperature that can be achieved by chemical means? The surface of the sun is about 10,000°F. Can that temperature be achieved on earth? What is the temperature of a nuclear explosion?

In addition to designating stars as type B, A, F, G, K, and M, astronomers also use the letters O, R, N, and S. What are the characteristics of each of these types of stars? Which ones are the hottest and which one the coolest? What are carbon stars? Why are some stars called diamond stars?

Extrasolar Planets

An extrasolar planet is one located outside our solar system and around another star. The difference in a star and planet is that a star produces heat and light while a planet shines by reflected light. Stars emit energy by nuclear fusion. During fusion (the word means "to melt together"), four hydrogen atoms are changed into one helium atom. A slight amount of mass is lost, and the mass is converted to energy.

The center of the sun is a place of intense pressure and high temperature, about 30 million degrees Fahrenheit. Like the sun, fusion takes place at the center of a star. The star has to be big enough to cause high pressure and heat. Astronomers have calculated that a body must be at least 13 times larger than Jupiter to start the nuclear fires burning. Any body smaller than 13 times the mass of Jupiter that orbits a star is a planet. However, detecting a planet — even a large one — around a distant star is a difficult task.

Planets are many times smaller than a star. In terms of stellar distances, planets

Explore

1. What is the zone of life?

2. What is a hot-Jupiter?

3. Have extrasolar planets been discovered?

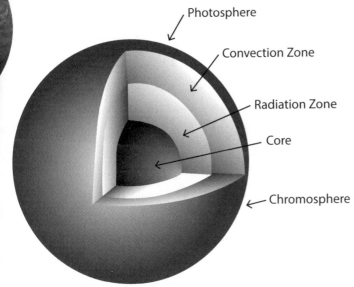

Photosphere

Convection Zone

Radiation Zone

Core

Chromosphere

are located very close to their parent star. A star is bright; the planet is close to the star and dim by comparison. This makes it difficult to directly detect a planet with a telescope.

One way to detect an extrasolar planet is by using the spectroscope. When a star is moving, the lines of the spectrum are shifted toward the red (going away from us) or toward the blue (coming toward us).

Sometimes a star has a double set of spectral lines constantly shifting back and forth. This shows that the star is actually binary — two stars in orbit of one another. Because they are orbiting their common center of

mass, one star is coming toward us while the other is moving the other way. This causes the spectroscope lines to double and to move back and forth.

Suppose, however, that instead of two stars, a star has a massive planet orbiting it. The planet produces no spectral lines. But the star and its planet move around the common center of mass. The star's spectral lines shift back and forth due to the Doppler effect. The star's spectral lines shift first toward the red as it recedes from the earth, then back toward the blue as it approaches the earth. The spectroscope can detect an extrasolar planet because of the shifting spectral lines.

The spectroscope method is also known as the radial velocity method. One advantage of the radial velocity method is that it not only reveals the existence of a planet, but also gives information about the mass of the planet. The more the planet affects the star, the greater the planet's mass.

The radial velocity method works best if a large planet is very close to its star. The planet would

The radial velocity method for discovering planets around distant stars uses the spectroscope. As a distant exoplanet orbits the star, it introduces changes in the frequency (color) of the star's light.

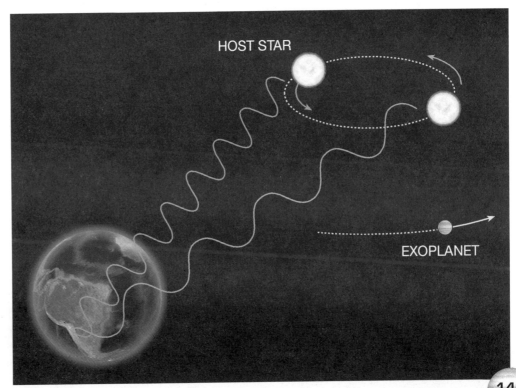

HOST STAR

EXOPLANET

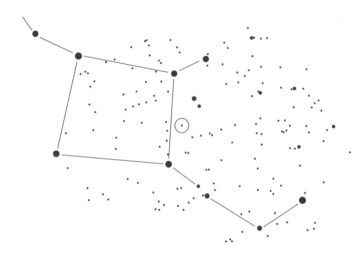

The discovery in 1995 of a planet near 51 Pegasi (circled) helped to ignite numerous efforts to find unknown planets using current methods of study, and by developing new ones.

quickly whip around the star and cause the star to produce very noticeable shifts in the spectral lines. Before 1995, astronomers expected planets the size of Jupiter to be far from their star and difficult to detect using the spectroscope.

However, the first extrasolar planet around a main sequence star was discovered using the radial-velocity method. In the constellation of Pegasus there is a fifth-magnitude star about 50 light-years away. The star is too dim to have a name, and is instead known as 51 Pegasi. It is of spectral type G. Our sun is also a spectral type G, and in many ways 51 Pegasi is similar to the sun.

The planet around 51 Pegasi was discovered using a sensitive spectroscope that could detect the slight and regular changes in the star's spectral lines. The rapid change in the spectral lines showed that the planet around 51 Pegasi was about half the mass of Jupiter and only 4.5 million miles from the star. The planet took only four earth days to make an orbit. In comparison, the sun's closest planet, Mercury, is 36 million miles from the sun and takes 88 days to make one complete orbit.

The discovery of the planet around 51 Pegasi was announced in October 1995. Two Swiss

astronomers, Michel Mayor and Didier Queloz, who worked at Haute-Provence Observatory in France, discovered it. The official name for the planet is 51 Pegasi B. The letter B shows that it was the first discovered orbiting its parent star. If more planets are found around 51 Pegasi, they will receive the letters C, D, E, and so on. The letter 'A' is reserved for the star the planets orbit.

The discovery of the 51 Pegasi B came as a surprise. Astronomers did not expect planets to be so close to its star, which made them much easier to detect. Suddenly, astronomers realized that planets could be found using current methods. The race was on to find more planets. Astronomers devoted more time to planet searches, and looked for additional ways to detect them.

A small planet farther away from a star would be more difficult to discover using the radial velocity method. But the planet's existence can be detected in other ways. The planet can eclipse its star and reduce the light reaching earth.

In 1782, John Goodricke, who was 17 at the time, reported that the second-brightest star in the constellation of Perseus varied in brightness. The young man was deaf and could not speak. But he could think, and he could communicate by writing. He submitted a paper to the Royal Society. He said that the variations in the star, known as Algol were regular. Algol's variation in apparent brightness

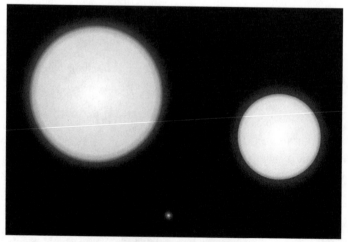

A size comparison between the white dwarf star Pegasi B (center), Pegasi A (left), and the sun (right)

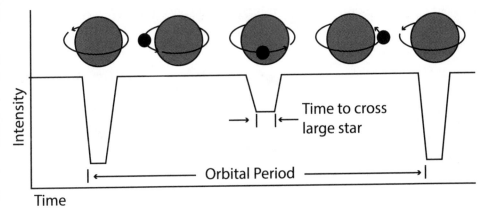

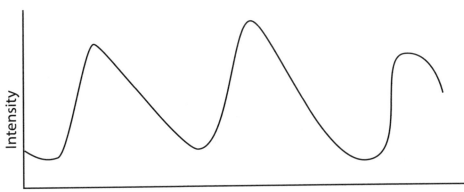

This diagram shows how the light curve of an eclipsing variable (top) is different than the light curve of a pulsating variable (bottom).

took place in 2 days, 20 hours, and 49 minutes. For most of that time it was second magnitude, the same brightness as Polaris, the Pole Star. For about four hours it faded to fourth magnitude and immediately began to brighten again.

John Goodricke suggested that an invisible companion star periodically eclipsed it. The star came in front of Algol and blocked part of its light. His suggestion turned out to be entirely correct. A dim companion star was eclipsing Algol.

The light curve of an eclipsing variable is unmistakable and quite unlike one of a pulsating variable such as a Cepheid. In an eclipsing variable, the light is steady until the dimmer star passes in front of the brighter one. Then the light dips until the dimmer one moves to the other side. Later, when the brighter star blocks light from the dimmer one, the light diminishes once again, but this time the dip is smaller.

A planet may also eclipse a star, reduce the light, and make its presence seen. The amount a planet dims the star depends on its size and the size of the star. The eclipsing method of planet detection is also known as the transit method. One benefit of the transit method is that it reveals the size of a planet. The greater the drop in brightness during the eclipse, the bigger the planet.

Some planets can be explored by both the radial velocity method using the spectroscope and the transit method using eclipses. The radial velocity method gives the mass of the planet, while the transit method gives its size. Once those two values are known, other information about the planet can be calculated, such as its density. A large planet of low density would probably be similar to Jupiter. A smaller planet of higher density would be more earth-like.

Using infrared light is another way to detect a planet. Infrared light has waves longer than red light, and it is invisible to the human eye. However, infrared light can be made visible with special equipment. When planets absorb light from a star, some is changed into heat. The heat energy is emitted as infrared light. The planets in our

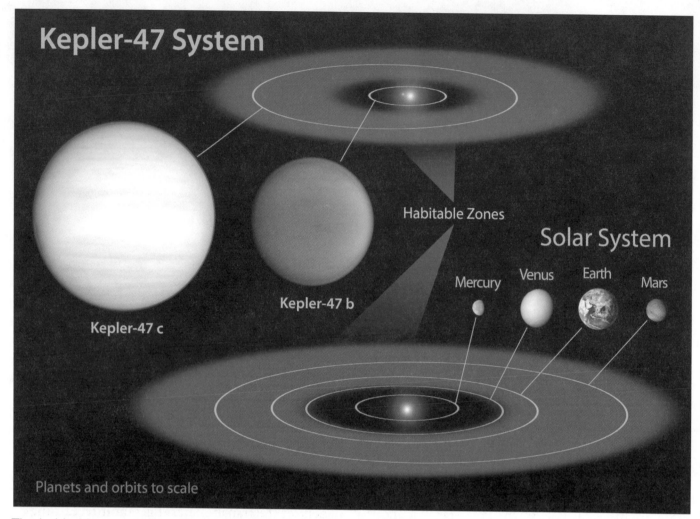

Kepler-47 System

Habitable Zones

Solar System

Kepler-47 c

Kepler-47 b

Mercury Venus Earth Mars

Planets and orbits to scale

The habitable zone is the area in a planetary system where liquid water might be found on the surface of a planet. Here we see a comparison of our own solar system's habitable zone to that of the Kepler-47 system. Kepler-47 contains two stars, neither of which is as bright as the sun.

solar system are relatively brighter in the infrared part of the spectrum than they are in the visible portion. For example, the sun is only 25 times brighter than Jupiter in the infrared, but 2.5 billion times brighter overall.

Planets around other stars are easier to see when the star is photographed in infrared rather than in visual light. Direct detection is one way to find extrasolar planets.

An extrasolar planet does not have to be seen directly for its presence to be revealed. Astronomers have been able to detect dim companion stars that orbit distant stars. Sirius was found to have a White Dwarf companion because Sirius appeared

to wobble. The White Dwarf and Sirius revolved around a common center of gravity. It pulled the brighter star off its straight-line motion.

The same wobble can also reveal the presence of a planet, especially if it were a large one like Jupiter. Strictly speaking, a planet does not orbit a star. Instead, star and its planet orbit their common center of mass. This is true of any two objects orbiting one another.

For example, we usually say that Jupiter orbits the sun. But that is not exactly correct. Both the sun and Jupiter orbit their common center of mass. If Jupiter and the sun were equal in mass, they would each circle around a point midway between

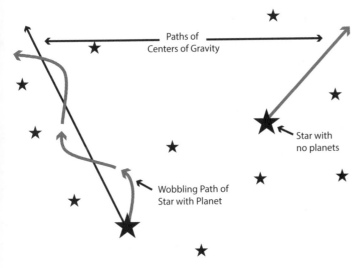

Paths of Centers of Gravity

Star with no planets

Wobbling Path of Star with Planet

the two. But Jupiter and the sun are not equal in mass. The sun is about 1,000 times more massive than Jupiter, so the balance point is much nearer the sun.

The center of mass of the Jupiter-sun system is about 465,000 miles away from the center of the sun. The sun is about 432,000 miles in radius. So the sun revolves around a point 33,000 miles above its surface. Jupiter gives the sun a very tiny, but distinct wobble. The larger the planet and the farther it is from its star, the greater a wobble it causes. The wobble that a planet causes in a star's motion is another potential way to detect extrasolar planets.

The first attempt to find an extrasolar planet was made using the wobble method. Starting in 1937, Peter van de Kamp made a detailed record of the motion of Barnard's star. During the next 30 years, he took 10,000 photographs through a telescope at the Sproul Observatory at Swarthmore College in Pennsylvania.

Barnard's star is named after Edward E. Barnard of the Yerkes Observatory. He detected the star in 1916. It had a noticeable amount of movement even from one year to the next. In the space of about 100 years his star would move across the sky by a little more than half the width of the full moon. Barnard's star is about six light-years away. It is the second-nearest star system to earth. Not only is it nearby, but it is a star of low mass, a small, red type-M star of the main sequence.

Because of its nearness and low mass, Barnard's star became a good candidate to be investigated for an extrasolar planet.

In 1964, Peter van de Kamp announced that the observations indicated that Barnard's star had a planet that was about half again as large as Jupiter. However, his work showed how difficult it is to detect a planet. It turned out to be an error. As instruments and methods used to measure his photographs improved, the amount of wobble grew less and less. If a planet does exist around Barnard's star, it would have to be far smaller than Jupiter. No planet of any size has been found around Barnard's star.

Although the wobble method is the easiest to understand, it is also the hardest to carry out. So far, only a couple of planets have been found using this method.

During the years from 1995 to 2012, almost 900 planets were detected. Of that number, about 300 planets were found using the transit (eclipse) method. About 500 were found by the radial velocity (spectroscope) method. Another 32 were directly photographed. All other methods account for about 50 planets.

An artist's conception of the red dwarf star CHXR 73 A and its companion object CHXR 73 B. The companion object is 12 times more massive than the planet Jupiter. CHXR 73 B may either be a planet, a failed star, or a brown dwarf.

A planet's position in relation to the star it may be orbiting can put it at risk. Here is an solar flare engulfing a planet

Human space travel to another star is out of the question today. Should it ever be possible, astronauts will most likely aim for a star that has a planet similar to earth. An earth-like planet would also need to be in a location around the star known as the zone of life: not too hot, not too cold. The zone of life is the orbit at a distance from a star where water is likely to be a liquid. The zone is not a single orbit around the star, but a band with a width that can vary depending on the star.

For our sun, the zone of life extends from about 75 million miles out to about 140 million miles. The earth is at 93 million miles from the sun right in the middle of the zone of life. Most of the surface of the earth is neither too hot nor too cold. Earth is the only planet in which water can be found in all three states: solid (ice), liquid (water), gas (clouds). All three exist at the same time in one place or another on earth. If earth were much closer to the sun, the water would boil and change into a vapor. If earth were much farther from the sun, the water would all be tied up in great frozen ice caps and glaciers.

Mars, at 140 million miles from the sun, just skims along at the outer edge of the zone. It has water, but almost completely in the form of ice. Venus, at 67 million miles from the sun, orbits too close to the sun. Venus has water, too, but in the form of steam. In our solar system, Venus and Mars just miss being in the zone of life.

Some stars can be eliminated right away as good choices for a stellar expedition from earth. For instance, variable stars are unlikely choices. Our sun is a star whose output of energy in the form of light and heat is practically constant. That's not the case for all stars. Some variable stars change by as much as a full magnitude. Others vary by as much as 10 magnitudes.

If the absolute brightness of the sun increased by half a magnitude, the average temperature on earth would increase by about 60 degrees Fahrenheit. If its intensity decreased by half a magnitude, the average temperature would fall by about the same amount. A variable star, then, is not a likely candidate for a visit by the first human expedition to a planet.

Red Giants are also poor choices. They have proven to be unstable. Sometimes they suddenly flare up in the form of a nova. A planet orbiting comfortably in the zone of life would suddenly find itself being washed over by a flare that would turn everything on its surface to a crisp.

A White Dwarf would also be a poor choice for a visit by explorers. Although its surface is about the same temperature as the sun, a White Dwarf is so small it radiates little energy. It would have an extremely narrow zone of life, and that zone of life would be very close to the star. Tidal effects would cause problems. The White Dwarf would produce tides on the planet that would act as a brake and bring the planet's rotation to a stop. For a planet to always keep the same side toward its star would be disastrous. One side would be baked with heat. The other side would radiate its heat into the dark of space and become bitterly cold. Water in the liquid state would be unlikely.

Planets have been found around binary stars. It would seem that planets in such a system would have a difficult time avoiding being too hot or too cold. Although computer models predict such suitable planets could exist, the first space explorers will be better served by finding a planet around a single star that is part of the main sequence.

Not all of the stars in the main sequence are suitable either. Recall that the main sequence is made of stars of type B, A, F, G, K, and M. The B and A stars are especially hot and bright. Stars of type G are like our sun. K and M stars are dim and cool.

For dim, red stars of the K and M type, the zone of life is close to the star and very narrow. It's like standing near a small fire. It's hard to find a comfortable location. Stars of class K and M have a much reduced zone of life. Planets are less likely to be found within the narrow zone in which water is maintained as a liquid.

In addition, planets around cooler K and M stars must be close to the star. Tidal effects, like those that would be a problem with White Dwarfs, would also be a problem with planets near K and M stars.

Another problem exists with K and M stars. All stars produce flares — unusually bright and hot tongues of gas ejected from their surfaces. On the sun, however, such an outburst adds only a small amount to the total solar output of light and heat. The same flare on an M star would increase its energy output by 50 percent.

Because of the narrow zone of life, tidal effects, and flares to which M stars are prone, they are unlikely to have a planet that would entice visitors from earth.

What about stars at the other extreme, stars of type B and A? These are hot, bright stars. They would have a broad zone of life. A planet is more likely to be found somewhere within the zone where water would be liquid. However, hot stars radiate a much larger fraction of their energy in the

Sometimes astronomers will take what they know about a possible planet that is discovered (position, composition, etc), and have an artist's illustration created like this one of PH1, discovered by volunteers of the Planet Hunters project. PH1 is a circumbinary planet in a four-star system, which means the planet orbits a double star that is being orbited by another pair of stars.

Our planet and our solar system were once considered to be rare in the universe. But in this artist's drawing, it shows how common planets orbiting stars are in the Milky Way. A six-year survey of millions of stars revealed that planets around stars are the interstellar rule rather than being a rare exception.

form of ultraviolet light and x-rays. Even our sun produces dangerous ultraviolet rays. An ozone layer in our atmosphere reduces it to a non-lethal level.

Planets of B and A stars would be bombarded by deadly radiations that even an ozone layer couldn't filter out. Of course, the planet might have some other more opaque layer in the atmosphere to block the rays, but that is unlikely.

Stars of type F, G, and K are the best candidates to have a zone of life that matches what we experience on earth. Some of the cooler K stars may have the same objections as M stars, and some of the hotter F stars may be too much like the B and A stars. But, overall, they are the best choices for looking for an earth-like planet within the zone of life.

Planets with the same mass and density as the earth have been found. Some even may have atmospheres. When they pass in front of the star, the light dims in such a way as to suggest that the planet has an atmosphere. However, they are not earth-like because none have been found in a zone of life.

From 1995 to 2012, not a single earth-like planet has been found in the zone of life

around a suitable F, G, or K type star. Astronomers predict that an earth-like planet in a zone of life will be found eventually. We should remember that even if they are found, earth is unique in our solar system. Elsewhere in the Milky Way, earth-like planets that are suitable for humans will be few and widely separated in the vastness of space.

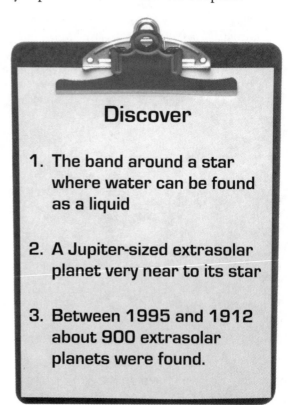

Discover

1. The band around a star where water can be found as a liquid

2. A Jupiter-sized extrasolar planet very near to its star

3. Between 1995 and 1912 about 900 extrasolar planets were found.

A B   1. Stars produce energy by the (A. fusion of hydrogen B. the fission of uranium).

A B C D   2. The first extrasolar planet was discovered by using (A. the radial velocity spectroscope method B. The transit method using eclipses C. looking in ultraviolet D. the wobble method).

T F   3. Algol was an eclipsing variable star discovered by John Goodricke who was deaf and could not speak.

A B   4. A small planet of high density is probably more nearly like (A. earth B. Jupiter).

A B   5. The greater the decrease in brightness of a star when eclipsed by a planet, the _____ the planet. (A. larger, B. smaller)

T F   6. Peter van de Kamp used the wobble method to conclusively prove that a planet orbited Barnard's star.

_____   7. In our solar system, _____ and Mars just miss being in the zone of life.

A B C D   8. Extrasolar planets suitable for human settlement are most likely to be found around (A. Red Giants and White Dwarfs B. stars of type O and B C. stars of type F, G, and K D. stars of type M).

T F   9. All stars produce flares.

A B C D   10. The number of earth-like extrasolar planets in the zone of life that were discovered between 1995 and 2012 number (A. 500 B. 300 C. 32 D. none).

EXPLORE MORE

Although lists of the top ten astronomers vary, certainly these individuals would be high on any list: Hipparchus of Nicaea, Claudius Ptolemy, Nicolaus Copernicus, Tycho Brahe, Galileo Gallilei, Johannes Kepler, Isaac Newton, Edmund Halley, William Herschel, and Edwin Hubble. Research each one, and write a single line that states the most important single achievement of each one. Example: Copernicus — proposed the sun-centered planetary system.

The skills for being an astronomer vary greatly. Some, such as Isaac Newton, were skillful at mathematics. Others, such as William Herschel, were patient observers. Edmund Halley could analyze existing information and propose new discoveries. Some, such as Henrietta Leavitt and Clyde Tombaugh, spent hours poring over photographs of stars. Astronomers study the sun, planets, stars, galaxies, and nebula. Think about your skills and abilities. What type of astronomy would you enjoy most? In which field do you think you could make useful discoveries?

Flares on the sun result in consequences here on earth. What are sunspots, and are they associated with flares? Research how flares cause aurora on earth and how they impact communication satellites. Do astronauts have to take special precautions following a solar flare eruption? What are the Van Allen Radiation Belts around earth? Who discovered them, and when were they discovered?

For the artist: Draw some other-world landscapes in color. Imagine sunrise over one of the moons of Jupiter; a planet orbiting close to a Red Giant; the flare of an M-type star washing over a nearby planet. What would earth's landscape look like if it and Mars switched places?

Starlighted Nights

Exploring the World of Astronomy has emphasized observational astronomy: what we can see with the unaided eyes, binoculars, and small telescopes. Known facts are emphasized. Before dwelling on theoretical discussion or speculation, it is best to first see what is actually visible. Nothing substitutes for observing the stars in the dark vault of the heavens. Surprise and delight often accompany the first good, clear view of a star cluster, contrasting double stars of blue or gold, or a misty nebula.

Probably the most important equipment for enjoying the night sky is not binoculars or a telescope. Instead, it is a reclining lawn chair, warm clothing in winter, and insect repellant in summer. Once the eyes become adjusted to the dark, the night sky becomes a rewarding show quite unlike anything else. Very likely at least one shooting star will be seen during a one-hour observing session. As meteors sweep across the sky, only the eyes can move fast enough to follow them.

Explore

1. What is the astronomical Royal Family?

2. Which season is the best one for observing Scorpius the Scorpion?

3. What celestial feature is the Goldfish?

Progress

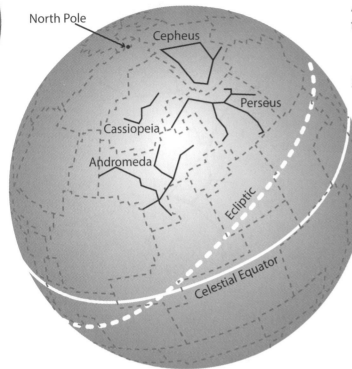

North Pole
Cepheus
Perseus
Cassiopeia
Andromeda
Ecliptic
Celestial Equator

Andromeda, Perseus, Cassiopeia, and Cepheus are the four constellations of the Royal Family.

Many people think interesting sky objects can only be seen in powerful telescopes. They are surprised to learn than the great galaxy in Andromeda (M31) is visible to the unaided eyes. It is, in fact, the most distant object that can be seen with eyes alone. M31 extends as a misty patch more than the apparent width of the full moon.

Binoculars and telescopes have their place, but for the beginning enthusiast, even an aimless sweep of the sky with the eyes can be enjoyable. Even more satisfaction is found in knowing what you are looking at and to be able to point out interesting objects. Learning to identify the planets, tracing out the constellations, and naming the principal and brightest stars is the place to begin.

A location without blindingly bright street lights or a strong city glow is better, but even in imperfect sky conditions, the brighter stars are visible. The outlines of most constellations are made by stars of fourth-magnitude or brighter. The dimmer fifth- and sixth-magnitude stars may be washed out. The outlines of constellations are still visible without the confusion of multitudes of fainter stars.

Ancient people must have felt a certain amount of awe when they saw the stars. Rather than being overwhelmed, they invented stories that gave the star patterns a more comfortable nature. They traced out starry outlines of animals, beasts, heroes, and villains. They turned the night sky into a parchment on which was written stories of suspense and adventure.

For instance, the Royal Family is made of four constellations: Andromeda, Perseus, Cassiopeia, and Cepheus. They are constellations of fall and are seen in the northern part of the sky. Andromeda is the daughter of Cassiopeia the queen and Cepheus the king. As the story goes, Andromeda is chained to a rock where she will fall victim to Cetus the Leviathan (sea monster). But Perseus, riding on his winged horse Pegasus, rescues her, and they are later married.

Another grouping is Orion the Hunter, his dogs Canis Major and Canis Minor, and Lepus the Hare, the rabbit they chase. When Orion throws his leg over the horizon in the east, there is a feeling of expectation because he brings the stars of winter: Betelgeuse and Rigel in Orion; Sirius in Canis Major; Canopus in Canis Minor; Aldebaran in Taurus the Bull; Castor and Pollux in Gemini the Twins. Right there, in a single sweep of the eyes are 7 of the 20 brightest stars.

Other easy-to-recognize patterns are the three summer constellations of Lyra the Harp, Aquila the Eagle, and Cygnus the Swan. They each

Scorpius the Scorpion

The Milk Dipper

have three first-magnitude stars — Vega, Altair, and Deneb that make the summer triangle.

In the brightest part of the Milky Way are Scorpius the Scorpion and Sagittarius the Archer. Scorpius is easy to recognize because of Antares, a bright, red star. Sagittarius has a smaller grouping of stars known as the Milk Dipper. It looks like a dipper that is scooping up part of the Milky Way.

For most hobbies, a guide or mentor is needed to become proficient in understanding what you are seeing. Astronomy is more fun if it is shared. However, the astronomy enthusiast can use one constellation as a steppingstone to nearby constellations. For instance, the Pointers (two stars in the front of the Big Dipper) point toward Polaris in Ursa Minor. The diagonal line of the stars in the bowl of the Dipper point to Castor, one of the two main stars in Gemini the Twins. The back two stars in the Dipper point to Regulus, the brightest star in Leo the Lion. Following the curve of the Big Dipper's handle takes you to Arcturus in Bootes the Herdsman.

The constellations are the background against which other objects such as planets

and the Messier objects can be found. One goal of the beginning astronomer is to learn the main constellations.

As a first step, learn to identify the circumpolar constellations. The advantage of the circumpolar constellations is that it doesn't make any difference what time of the year you start. The circumpolar constellations include Ursa Major that contains the Big Dipper, Ursa Minor that contains the Little Dipper, the long and winding Draco the Dragon, and the big W of Cassiopeia the Queen, and Cepheus the King.

Once away from the circumpolar constellations, you have to be aware of what is visible during each season. Because of the earth's motion around the sun, each season has different constellations of stars that are best placed for viewing.

Stars move east to west on a seasonal basis. Those stars visible at 9:00 in the middle of one month will be 30 degrees to the west when viewed at 9:00 in the middle of the next month. The year has 12 months, and the complete circle of the sky is 360 degrees. Dividing 360 by 12 gives 30 degrees. Each month represents a 30-degree shift in the stars.

When we switch from one season to the next, the stars that were favorably placed for viewing have been carried out of sight below the western horizon. The stars that would only be visible well after midnight have now become favorably placed in the evening sky.

Stars also move east to west on an hourly basis because of the daily rotation of the earth on its axis. Those stars visible at 9:00 on any night will be 30 degrees to the west when viewed two hours later at 11:00 on the same night. The day has 24 hours, and the complete circle of the sky is 360 degrees. Dividing 360 by 24 gives 15 degrees. Two hours of the daily rotation will change the location of stars by 30 degrees — the same extent as one month in the annual revolution. Later in the night, stars normally visible later in the year come into view. A longer viewing session will rotate a greater part of the sky into view.

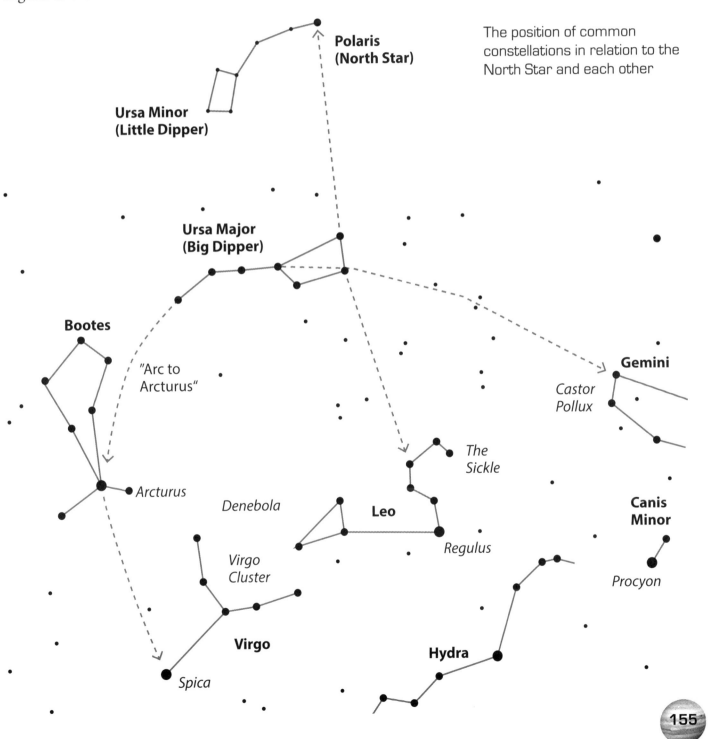

The position of common constellations in relation to the North Star and each other

Northern Hemisphere Constellations

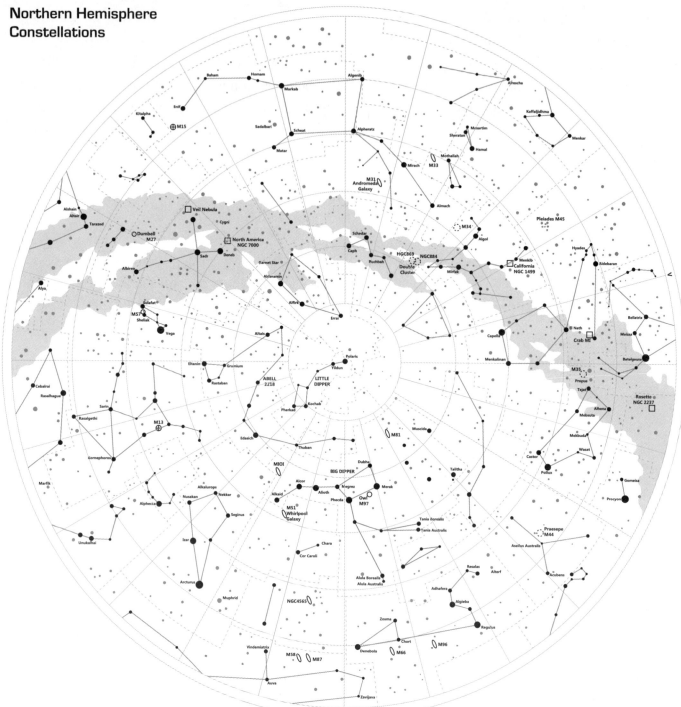

A star chart showing the location of many known distant space objects, nebulas, and constellations. Star charts are an important way for you to find what you are looking for in the night sky. For example, to find the Whirlpool Galaxy (M51), you can use the stars of the Big Dipper's handle to find its position.

North Polar Constellations

Camelopardus	Draco
Cassiopeia	Ursa Major
Cepheus	Ursa Minor

South Polar Constellations

Apus	Norma
Chamaeleon	Octans
Circinus	Pavo
Crux	Triangulum
Dorado	Australe
Hydrus	Tucana
Mensa	Volans
Musca	

Northern Hemisphere Autumn (Southern Hemisphere Spring)

Andromeda	Perseus
Aquarius	Phoenix
Aries	Piscis Austrinus
Cetus	Pisces
Grus	Sculptor
Lacerta	Triangulum
Pegasus	

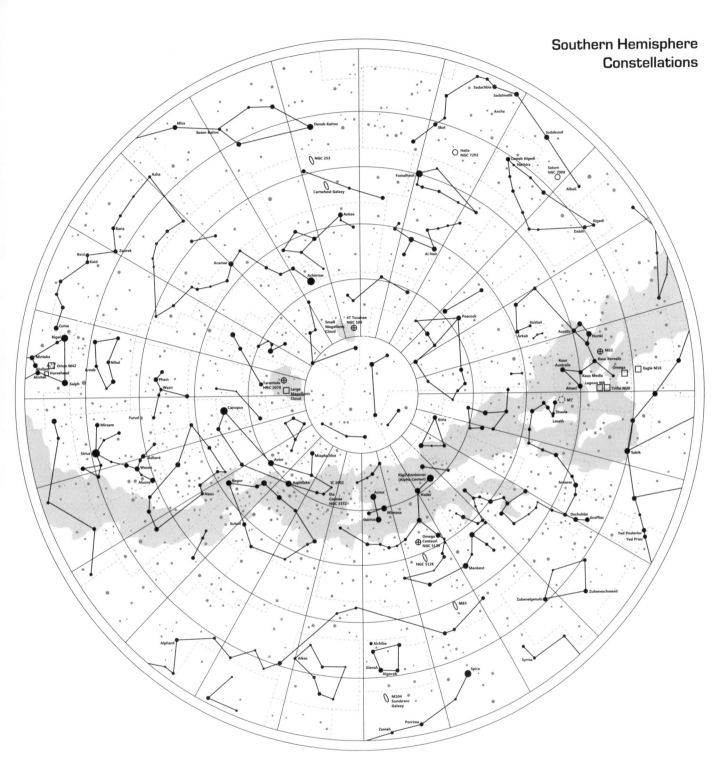

Northern Hemisphere Winter
(Southern Hemisphere Summer)

Auriga	Monoceros
Caelum	Orion
Canis Major	Pictor
Canis Minor	Puppis
Carina	Reticulum
Columba	Taurus
Eridanus	Vela
Fornax	
Gemini	
Horologium	
Lepus	

Northern Hemisphere Spring
(Southern Hemisphere Autumn)

Antlia	Lupus
Boötes	Lynx
Cancer	Pyxis
Canes Venatici	Sextans
Centaurus	Virgo
Coma Berenices	
Corvus	
Crater	
Hydra	
Leo	
Leo Minor	

Northern Hemisphere Summer
(Southern Hemisphere Winter)

Aquila	Lyra
Ara	Microscopium
Capricornus	Ophiuchus
Corona Australis	Scorpius
Corona Borealis	Scutum
Cygnus	Serpens
Hercules	Sagitta
Delphinus	Sagittarius
Equuleus	Telescopium
Indus	Vulpecula
Libra	

The best constellations to learn after the circumpolar ones are those in the ecliptic. The ecliptic is the flat surface, or plane, that contains the orbit of the earth around the sun. When the plane is projected onto the sky, it passes through the constellations of the ecliptic. Because the other planets travel very nearly in the same plane as the earth, they will be found close to the ecliptic. Knowing the constellations of the ecliptic makes finding the planets easier.

The ecliptic comes from the word eclipse. The sun and moon must be in the plane of the ecliptic for a lunar or solar eclipse to be visible from earth. If either one is not in the plane of the ecliptic, then the moon will pass above or below the sun, and an eclipse will not occur.

The ecliptic is sometimes confused with the zodiac. Astrologers use the signs of the zodiac to cast a horoscope — a prediction of whether a person will have good luck or bad luck based on the zodiac sign under when the person was born and the position of the stars and planets. Astrology is a dreadful suppression. It should not be confused with the true science of astronomy.

Astronomers speak of the constellations of the ecliptic; astrologers talk about the signs of the zodiac. The constellation Scorpius is not spelled the same as the zodiac sign Scorpio, and the constellation Capricornus the Goat of the ecliptic is not spelled the same as the zodiac sign Capricorn. In addition, the zodiac has 12 signs; the ecliptic has 13 constellations. The 13th ecliptic constellation is Ophiuchus the Snake Holder.

Another useful line for finding the constellations is the line across the sky that goes directly north to south. For observers in the Northern Hemisphere, astronomers have calculated when a constellation will be best viewed at 9:00 p.m. with that line passing through them. The line is called a meridian.

Look at table Constellations of the Ecliptic. Each constellation has a date in which it is located on the north-south meridian line at 9:00 p.m. For instance, to find Aries the Ram, face south at 9:00 p.m. on December 10. It will be directly along the line running north to south. It is one of the dimmer and more difficult constellations to find, so knowing exactly where to look is great help. To the east is Taurus the Bull with its distinctive V-shaped horns. Taurus crosses the 9:00 p.m. meridian on January 15. East of Taurus is Gemini the Twins, and they are visible along the 9:00 p.m. meridian on February 20.

Constellations of the Ecliptic	
Constellation	Date of 9:00 p.m. meridian crossing
Aries the Ram	December 10
Taurus the Bull	January 15
Gemini the Twins	February 20
Cancer the Crab	March 15
Leo the Lion	April 10
Virgo the Virgin	May 25
Libra the Scales	June 20
Scorpius the Scorpion	July 20
Ophiuchus the Snake Holder	July 25
Sagittarius the Archer	August 20
Capricornus the Goat	September 20
Aquarius the Water-bearer	October 10
Pisces the Fishes	November 10

The sun followed the ecliptic. In winter the sun is lower in the sky and in summer higher in the sky.

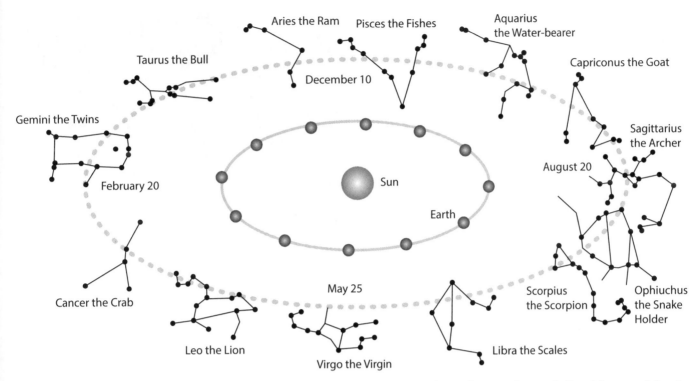

During the year, different constellations of the ecliptic are visible along the north-south line (the meridian) at 9:00 pm.

The listing of constellations in the table begins with the three winter constellations: Aries, Taurus, and Gemini. Next are the three constellations of spring: Cancer, Leo, and Virgo. Summer is the only season with four constellations in the ecliptic: Libra, Scorpius, Ophiuchus, and Sagittarius. The final three are the constellations of autumn: Capricornus, Aquarius, and Pisces.

The line of the ecliptic sinks down toward the south during the summer months but is higher in the sky during winter. The variation in its location is caused by the tilt of the earth's axis. During winter, the axis of the earth is tilted toward the ecliptic, making it higher in the Northern Hemisphere. During summer, the axis of earth is tilted away from the ecliptic-making it lower and more nearly along the southern horizon.

The tilt puts two of the summer constellations of the elliptic very close to the southern horizon for observers in the northern United States. It may be difficult to see the entire sweep of Scorpius the Scorpion and Sagittarius the Archer.

In 1929, the International Astronomical Union officially defined 88 constellations across the span of the sky. The ancient Greeks named many of the constellations of the Northern Hemisphere more than 2,000 years ago. Astronomers retained those names. However, many of the southern constellations were named in more recent times. They

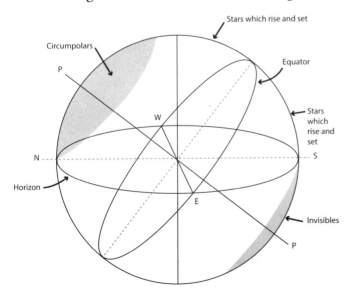

We can see stars from the Southern Hemisphere rise and set every night. The diagram shows this progression from a point on the latitude of 45° N.

include Dorado the Goldfish, Horologium the Clock, Microscopium the Microscope, Pyxis the Compass, and even Telescopium the Telescope.

It is possible to see some of the constellations of the Southern Hemisphere from the Northern Hemisphere. It depends upon the observer's latitude — distance north or south of the equator. An observer at the North Pole would only be able to see Northern Hemisphere constellations. However, as one travels south, more and more of the southern constellations come into view. At the equator, all 88 of the constellations would be visible, although those near the North Celestial Pole and South Celestial Pole would be right on the horizon.

Find your approximate latitude to learn how far south you can see. Subtract that number from 90, and the result is how far south of the equator you can see. In the United States, a northern city such as Minneapolis, Minnesota, is at about 45° north latitude. From that location stars as far south as 45° south latitude can be seen. Observers in St. Louis, Missouri, which is at 39° north latitude, can see another six degrees farther south. A few southern United States locations such as Key West, Florida, or Brownsville, Texas, are at about 25° north latitude. From that latitude line, the best-known southern constellation can be seen — Centaurus the Centaur. Centaurus is home to the closest stars to the earth (except for the sun itself, of course).

As one becomes knowledgeable about the constellations, it is possible to recognize even a small portion of the constellation as it rises in the east.

The American poet Ralph Waldo Emerson wondered what would happen if the stars appeared only one night in a thousand years. Those who witnessed the event would treasure the sight. They would tell the generations that followed about that starlighted night. For us, the wonder and glory of the heavens are not 1,000 years away. Instead, they are put on display every clear night of the year.

Enjoying the night sky can be more than just an educational venture – it can become a great family adventure!

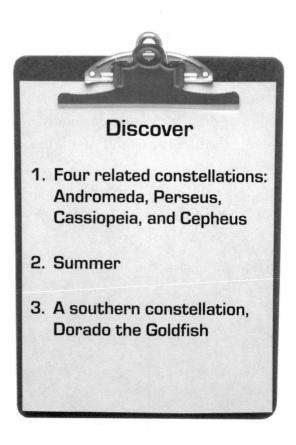

Discover

1. **Four related constellations: Andromeda, Perseus, Cassiopeia, and Cepheus**

2. **Summer**

3. **A southern constellation, Dorado the Goldfish**

A B C   1. *Exploring the World of Astronomy* has emphasized (A. astrophysics B. observational astronomy C. theoretical astronomy).

A B C D   2. To learn the constellations, first learn (A. the circumpolar ones B. those near Orion C. the Royal Family D. the Summer Triangle).

A B   3. Each month the stars move 30 degrees to the _____ when viewed at the same time each night. (A. west B. east)

T F   4. The word ecliptic comes from the word eclipse.

A B   5. The ecliptic is a plane that contains the (A. average orbit of all planets B. orbit of earth).

A B   6. The true science is (A. astronomy B. astrology).

A B C D   7. Refer to the table Constellations of the Ecliptic: Aries the Ram is a constellation of (A. summer B. autumn C. winter D spring).

T F   8. The Southern Hemisphere sky has an official constellation named Telescopium the Telescope.

T F   9. Southern constellations can only be seen from locations south of the equator.

EXPLORE MORE

What is Greenwich Mean Time? What is Zulu time? How many time zones are in the United States? How many in the world? St. Louis, Missouri, is almost exactly at 90 degrees west latitude. When a particular star is along the 9:00 p.m. meridian at Greenwich (0° longitude), how many hours will pass before the star is directly along the same line in St. Louis, Missouri?

For the artist: Draw figures to represent the animal constellations in the ecliptic: Aries the Ram, Taurus the Bull, Cancer the Crab, Leo the Lion, Scorpius the Scorpion, Capricornus the Goat, and Pisces the Fish.

The rotation of the earth that causes the stars to move west to east also causes the sun to appear to move during the day. Ancient people used the sun and stars to measure time. Research some of the forms sundials have taken to measure the time of day. Find directions for making a simple one from heavy paper and build your own sundial.

How many items have you added to your astronomy journal of objects seen? How many circumpolar constellations have you identified? How many constellations of the ecliptic? Have you succeeded in finding constellations that are in the Southern Hemisphere?

Explore More

CHAPTER 1

Many of the prehistoric sites such as Stonehenge in England, Chichen Itza in Mexico, and Casa Grande in Arizona use the sun and moon to mark seasons. Research one of these ancient observatories and describe some of the other astronomical events they could predict.

What did scientists learn about the composition of the moon rocks returned to earth by the astronauts?

CHAPTER 2

Research how to fully describe the elliptical orbit of a planet such as Mars. Draw an ellipse and label the semi-major axis, semi-minor axis, foci, eccentricity, inclination, ascending node, descending node, reference plane, apogee, and perigee.

Jules Verne lived in the late 1800s and early 1900s and wrote what today would be called science fiction, although at the time he called them "extraordinary voyages." H.G. Wells lived at about the same time. Give short summaries for some of the books that Verne and Wells wrote dealing with astronomical subjects.

View a photograph of Mars from a distance so it appears about the size of a dime held at arm's length. Sketch what you see.

What is terraforming? How could Mars be terraformed? What would be some of the reasons for and against terraforming Mars?

CHAPTER 3

Climbing Mount Everest on earth and surviving at its summit requires special equipment that would be similar to surviving on the surface of Mars. What additional challenges do travelers face when going to Mars and surviving on its surface?

Begin a journal detailing the planets and sky objects you have seen. List the circumstances under which you glimpsed the object: time of day or night, cloud conditions, sky conditions (clear, still air, do the stars shine with a steady light or is there turbulences causing them to twinkle), distance of object above the horizon, whether streetlights or city glow interfered, whether seen with unaided eyes, binoculars, telescope, etc.

CHAPTER 4

List the diameters of the following objects and rank them in size from the smallest to the largest: the four satellites of Jupiter, the planets Mercury, Venus, Earth, Earth's moon, and Mars. Which ones are the largest and the smallest? Which one is most nearly the same size as Mercury? Which one is most nearly the same size as the earth's moon?

Over several nights observe the moons of Jupiter with a small telescope. A spotting scope or 10x binoculars will also work — either one will need to be mounted so as not to jiggle. Sketch the positions of the visible satellites as Galileo did. Compare your sketches with their positions as published on the Internet. (One source is http://www.skyandtelescope.com.) Which satellites did you see most easily? Was one too faint or too poorly positioned for you to see?

Explore More

CHAPTER 5

For the Math Whiz: Compare the speeds of Mercury and Saturn in their orbits. Use the fact that the radius of Mercury's orbit is about 36,000,000 miles and it takes 88 days to orbit the sun, calculate its speed in orbit in miles per hour. Calculate the circumference of the orbit $C = 2\pi r$ and then divide by the number of hours in 88 days. Next, given that Saturn is 893,000,000 miles from the sun and takes 29.5 earth years to complete one orbit, calculate its speed in miles per hour.

Continue your journal of astronomical observations. Which of the following have you seen: Mercury, Venus, Earth, Earth's moon, Mars, Jupiter, Jupiter's four larger satellites, Saturn, Saturn's moon Titian. Have you seen a solar or lunar eclipse?

CHAPTER 6

Using an appropriate scale for the size of paper you have, make nested circles to scale of the Jovian Planets. Label each circle with the planet's name and color each circle to match the color of the planet.

In what fields do amateur astronomers work with professional astronomers today? What have amateurs contributed to astronomy?

For the math whiz: Compare the diameter of Uranus (31,763) and earth (7,926 miles). How many times larger is Uranus than earth? Calculate their surface area ($A = 4\pi r^2$, and $r = \frac{1}{2} D$.) How many times greater is the surface area of Uranus? Do the same for volume ($V = 4/3\pi r^3$.)

Generate a project of your own: make a model, color a poster, make a collage, write a poem, write a science-fiction story, interview a person who has been an amateur or professional astronomer, or choose some other interesting way to explore more.

CHAPTER 7

Read about one of the asteroid strikes on earth such as the one that created the Barringer Meteor Crater in Arizona, the Tunguska event in the uninhabited part of Siberia, Russia, in 1908, or the damage done by the meteor that exploded over the region of Chelyabinsk, Russia, in 2013. How extensive was the damage? What steps are being taken to prevent meteor and asteroid strikes from causing widespread damage on earth?

Look up the predicted times of meteor showers and meteor storms. How many known meteor showers are there? Currently, which one is predicted to be the strongest? What is a bolide?

Although Ceres is the largest asteroid and the first one discovered, it is not the brightest. Which asteroid is the brightest? What is its maximum magnitude? Why do scientists think it has a higher reflectivity (albedo) than the other large asteroids?

What have been the brightest comets in the last 100 years? When is Halley's comet expected to return? Do astronomers believe it will be as bright the next time as it was in the previous passages?

Explore More

CHAPTER 8

The magnification of an amateur telescope is given by the focal length of the main lens or mirror divided by the focal length of the eyepiece. An amateur 8-inch telescope with a focal length of 64 inches has eyepieces of 1 inch, ½ inch, and ¼ inch. What are the three magnifications possible with that telescope?

Astronomical telescopes give an upside-down view. How can the upside-down view be changed so an astronomical telescope can be used for viewing terrestrial subjects? How do spotting scopes and binoculars give a right-side-up view?

What is an alt-azimuth mounting for a telescope tube? What is an equatorial mount? What are the advantages and disadvantages of each? Which is best for long exposure photography? Why must the clock controlling the telescope for long exposure photography be set to sidereal time?

CHAPTER 9

What is the Royal Society? How did it get started? Who are some of the early members of the organization? What was Robert Hooke's role in that organization? Why is prompt communication of new discoveries important? What is a peer review?

What is the difference between applied science and theoretical science? In what ways can astronomy be applied? In what ways is it theoretical? Experiments impossible to duplicate on earth occur in outer space. Astronomers and astrophysicists can see the effects of crushing gravity, intense radiation, powerful magnetic fields, and extremes of heat and cold. In what ways do you think the discoveries of astronomy could have a benefit in daily life? Is it necessary for science to have an immediate benefit for it to be worthy of pursuit?

CHAPTER 10

Many state and country flags have stars on their flags. Do any states or countries have stars that represent actual star groupings or constellations?

From a list of the 110 Messer objects, make a table showing the number of each type: open clusters, globular clusters, galaxies, planetary nebula, etc. How are clusters, nebulas, and galaxies different? Some have an overall magnitude brighter than magnitude six. Why, if their magnitude is brighter than sixth magnitude, are they difficult to see with the eyes alone?

CHAPTER 11

What is the Doppler effect? How did Christian Doppler measure it? What applications are made of the Doppler effect today? What is Doppler radar? Why is it more useful than regular radar in measuring rainfall and winds?

CHAPTER 12

For several years, Peter van De Kamp believed he had discovered a planet around Barnard's star. Unfortunately, as time passed, his results were called into question. Research his work and discover why astronomers now discount his claim to have discovered the first extrasolar planet.

Questions

WHOLE BOOK REVIEW

_____ 1. Astronauts first landed on the moon in the Sea of _____.

_____ 2. The brightest planet is _____.

A B C D 3. Vesta, Pallas, Juno, and Ceres are names of (A. asteroids B. comets C. extrasolar planets D. Plutoids).

_____ 4. The largest planet is _____.

_____ 5. Callisto, Europa, Ganymede, and Io are names of the satellites of _____.

A B C D 6. The rings of Saturn are made of (A. gas and dust B. liquid water C. small, solid particles D. a solid sheet of metal).

T F 7. William Herschel discovered Uranus with a telescope he had built himself.

_____ 8. The Jovian planet discovered by pen and paper calculations before it was seen in a telescope was _____.

_____ 9. Comet material can be compared to a _____ snowball.

_____ 10. The Crab Nebula is the remains of a _____.

A B 11. The type of telescope that uses a mirror to focus light is a (A. reflector B. refractor).

_____ 12. The instrument that allows astronomers to learn the composition of a star and measure its speed is the _____.

A B C D 13. The scientist who showed that planets travel in elliptical orbits was (A. Tycho B. Kepler C. Galileo D. Newton).

A B C 14. The best way to view shooting stars is with (A. the eyes alone B. binoculars C. telescope).

Index

Resources

How to select a guide to the stars: Many field guides and Internet resources show constellations. The best ones represent constellations as simple line drawings that connect stars of 1st through 4th magnitude, but do not become cluttered with dimmer stars. Constellations should be grouped by seasons and show when they will be visible at some set time, such as 9:00 p.m.

Two historical books that remain viable as reference guides:

Garrett P. Serviss, *Astronomy with an Opera-Glass* (New York: D. Appleton and Company, 1888). Mr. Serviss's opera-glass had a magnification of about 4x. Subjects described in this book are readily visible in binoculars, which generally range in power from 6x to 10x. The book was highly popular when first published and was printed on good quality paper and with a strong binding, so many still exist. Copies are available at reasonable cost from Internet used bookstores.

Rev. T.W. Webb (edited by Margaret W. Mayall), *Celestial Objects for Common Telescopes* (New York: Dover Publications, 1962.) First published in 1859, the book contains a list of nebulas, star clusters, variable stars, etc., grouped by constellations. The common telescope that Rev. Webb used had a maximum usable magnification of 200x, so all of the objects described in the book are well within reach of the typical amateur telescope. *Volume I, The Solar System*, has not been revised but does describe the visual appearance of the planets. *Volume II, The Stars*, was revised by Margaret W. Mayall in 1962. Both books are readily available from Internet used bookstores at minimal cost.

Internet Resources: Websites can change. It is recommended parents confirm that these websites are appropriate for those who use them.

http://www.astronomy.com/ — The on-line website by *Astronomy* magazine. See also the *Sky and Telescope* website.

http://www.hubblesite.org/ — All about the Hubble Space Telescope including a gallery of photographs.

http://www.nasa.gov/ — Photos and the latest news about NASA space probes exploring the planets.

http://www.planetary.org/ — A volunteer organization for promoting sky observing and the exploration of space.

http://www.skyandtelescope.com/ — The on-line site of *Sky and Telescope* magazine. Includes a listing of what is visible each week in the night sky.

Most observatories such as Kitt Peak National Observatory, Palomar Observatory, and Keck Observatory on Mauna Kea in Hawaii have websites that provide material of interest to amateur astronomers.

Answers

Chapter 1 – Exploring the Moon

1. false — craters can also be seen.
2. B. — lava
3. true
4. B. — successfully landed
5. B. — one-fourth
6. false — about the same size
7. true
8. total
9. A — large
10. C — 29.5 days
11. harvest
12. true

Chapter 2 – Mars

1. wandering
2. C — Saturn
3. true
4. A — Copernicus
5. false — parallax of stars could not be seen.
6. B — Syrtis Major
7. A — close
8. Mars
9. False — *Phobos* means fear and *Deimos* means panic
10. B — small
11. D — wife
12. A — Percival Lowell
13. false — it had no artificial waterways.
14. C — Jonathan Swift

Chapter 3 – Terrestrial Planets

1. Mercury, Venus, Earth, Mars
2. true
3. false — Mars has larger canyons
4. A — carbon dioxide
5. A — iron oxide
6. B — an elliptical
7. A — Earth
8. Venus
9. true
10. phases
11. B — Venus
12. B — high temperature
13. A — 88 days
14. true

Chapter 4 – Jupiter

1. true
2. square
3. B — Galileo
4. D — Titan, the fourth is Io
5. B — Mercury
6. to match the inverted view of astronomical telescopes
7. false — more quickly
8. true
9. time
10. B — longitude — distance east or west
11. A — earth
12. square
13. D — the time for light to travel from Jupiter to earth
14. *Voyager*

Chapter 5 – Saturn

1. true
2. C — Saturn
3. Sirius
4. Titian
5. C — Christiaan Huygens
6. B — inclination
7. a gap in Saturn's rings
8. A — least
9. C — small, solid particles
10. false — without breaking up
11. A — shepherd
12. methane
13. helium
14. false — they know more about Titan

Milky Way over the desert of Bardenas, Spain

The iconic Horsehead Nebula is a favorite target for amateur and professional astronomers. It is shadowy in optical light. It appears transparent and ethereal when seen at infrared wavelengths.

Chapter 6 – Jovian Planets

1. hydrogen
2. false — as a music teacher
3. D — that he rented
4. B — stars
5. D — all of the above
6. true
7. D — Titania
8. comets
9. B — chief wrangler
10. D — Uranus
11. true
12. C — Johann Galle in Germany
13. true
14. Neptune
15. A — less

Chapter 7 – Plutoids and Denizens of Space

1. C — plutoids
2. D — Percival Lowell
3. blink
4. true
5. false — Pluto has at least five satellites
6. Neptune
7. A — Karl Gauss
8. false — between Mars and Jupiter
9. A — asteroids
10. true
11. true
12. storms
13. B — comets
14. snowball
15. A — they grow weaker

Chapter 8 – Telescopes

1. false — he built one based on a description of the telescope
2. prism
3. B — refractor
4. A — long
5. true
6. C — William Parsons, Lord Ross
7. true
8. gather
9. false — the 40-inch Yerkes refracting telescope is still the largest in the world
10. A — bubbles and strains in the glass
11. D — on Mt. Wilson
12. B — light pollution
13. false — Keck I and II are larger
14. A — with a serious design flaw

Chapter 9 – Breakthroughs in Astronomy

1. D — Isaac Newton
2. False — Tycho lived before the telescope was invented
3. A — Mars
4. true
5. B — Galileo
6. true
7. B — Christopher Wren
8. B — precisely locate the position of a network of stars
9. D — Ferdinand Magellan
10. comets
11. false — Sirius not Polaris
12. A — Fredrich Bessel
13. Centauri

Chapter 10 – Deep Sky Wonders

1. false — most are second magnitude
2. A — 30 minutes
3. A — an asterism
4. B — Ursa Minor
5. A — binary
6. A — circumpolar
7. D — open cluster
8. false — about 100 times brighter
9. true
10. comet
11. C — Omega
12. A — dust and gas blocked more distant stars
13. C — by plotting the positions of globular clusters

Chapter 11 – Stars

1. helium
2. 10,000
3. A — hotter
4. M
5. False — they are separate from the main sequence
6. true
7. A — an interferometer
8. A — Red Giants
9. False — Alvan Clark
10. A — more
11. B — long-term variable
12. C — Henrietta Leavitt
13. B — red
14. true

Chapter 12 – Extrasolar Planets

1. A — fusion of hydrogen
2. A — the radial velocity spectroscope method
3. true
4. A — earth
5. A — larger
6. False — his results are discounted today
7. Venus
8. C — stars of type F, G, and K
9. true
10. D — none

Chapter 13 – Starlighted Nights

1. B — observational
2. A — the circumpolar ones
3. A — west
4. true
5. B — the orbit of earth
6. A — astronomy
7. C — winter
8. true
9. false — some can be seen from northern latitudes.

Whole Book review

1. Tranquility
2. Venus
3. A — asteroids
4. Jupiter
5. Jupiter
6. C — small, solid particles
7. true
8. Neptune
9. dirty
10. supernova
11. A — reflector
12. spectroscope
13. B — Kepler
14. A — the eyes alone

Ius Chasm, a geologic feature found within the vast Valles Marineris on Mars; it is believed to have formed from the process of sapping, when groundwater erodes soil from a slope, much like it does in the process of forming gullies.

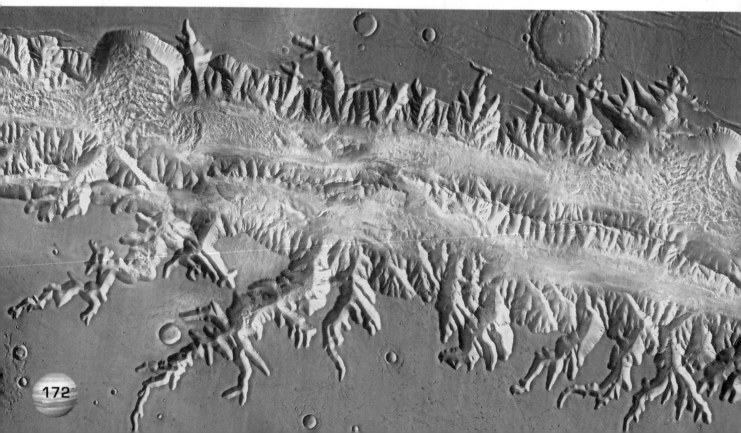

About the Author

John Hudson Tiner has a long-time interest and broad knowledge of astronomical subjects. He has studied on the graduate level in astronomy at Sam Houston State University, taught astronomy to high school students, and served as newsletter editor for a group of amateur astronomers. He has built several telescopes and used them for observing many of the sights described in this book.

In addition, he and his wife, Jeanene, have visited world-class observatories such as Yerkes Observatory in Williams Bay, Wisconsin, Kit Peak National Observatory in Arizona, Palomar Observatory in California, and Mauna Kea Observatory in Hawaii. They also traveled to Birr Castle Ireland, site of the Leviathan, the largest telescope in the world in the 1800s. Other astronomical related sites they have visited include the Arecibo Radio Telescope in Puerto Rico, the Very Large Array in New Mexico, and the world's largest fully steerable radio telescope at Green Bank, West Virginia. They have visited prehistory sites such as Chichen Itza, Mexico; Medicine Wheel, Wyoming; Casa Grande, Arizona; and the standing stones of Scotland.

John Hudson Tiner is the author of textbooks, science curriculum material, character-building biographies, and books on a variety of other subjects. His books entertain while they increase a person's knowledge of the essentials of the subject. He has a master's degree from Duke University. He brings a wide range of exciting facts together to write more than 1,000 published manuscripts, including 80 books, for all age groups.

He is well known for his books for teens and young people, including the "Exploring the World of" series for Master Books. He says, "Young readers of today deserve to experience the thrill and wonder of learning about science. They delight in learning new facts, especially if the facts are presented in an exciting way."

His material has been translated into a variety of languages, including Spanish, Russian, Indonesian, and German.

From the Center of the Sun to the Edge of God's Universe

Think you know all there is to know about our solar system? You might be surprised!

Master Books is excited to announce the latest masterpiece in the extremely popular *Exploring Series, The World of Astronomy*. Over 150,000 copies of the *Exploring Series* have been sold to date, and this new addition is sure to increase that number significantly!

- Discover how to find constellations like the Royal Family group or those near Orion the Hunter from season to season throughout the year.
- How to use the Sea of Crises as your guidepost for further explorations on the moon's surface
- Investigate deep sky wonders, extra solar planets, and beyond as God's creation comes alive!

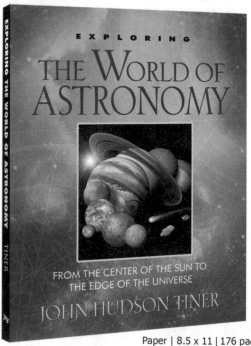

Paper | 8.5 x 11 | 176 pages
978-0-89051-787-1

The book includes discussion ideas, questions, and research opportunities to help expand this great resource on observational astronomy.

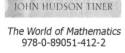
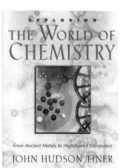
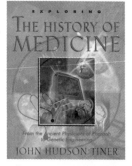

JACOB'S MATH

ELEMENTARY ALGEBRA

This high school algebra curriculum provides a full year of math in a clearly written format with guidance for teachers as well as for students who are self-directed. The full-color student text is divided into 17 sections, covering functions and graphs, integers, rational numbers, exponents, polynomials, factoring, fractions, and more.

978-1-68344-263-9

GEOMETRY

Harold Jacobs' *Geometry* has been an authoritative standard for years, with nearly one million students having learned geometry principles through the text. With the use of innovative discussions, cartoons, anecdotes, and vivid exercises, students will not only learn but will also find their interest growing with each lesson.

978-1-68344-260-8